05-CTS-3

EVALUATION OF PROCESSES TO REDUCE ACTIVATED SLUDGE SOLIDS GENERATION AND DISPOSAL

by:
Julian Sandino
Drury Whitlock
CH2M HILL Canada Limited

2010

WERF

The Water Environment Research Foundation, a not-for-profit organization, funds and manages water quality research for its subscribers through a diverse public-private partnership between municipal utilities, corporations, academia, industry, and the federal government. WERF subscribers include municipal and regional water and wastewater utilities, industrial corporations, environmental engineering firms, and others that share a commitment to cost-effective water quality solutions. WERF is dedicated to advancing science and technology addressing water quality issues as they impact water resources, the atmosphere, the lands, and quality of life.

For more information, contact:
Water Environment Research Foundation
635 Slaters Lane, Suite G-110
Alexandria, VA 22314-1177
Tel: (571) 384-2100
Fax: (703) 299-0742
www.werf.org
werf@werf.org

This report was co-published by the following organization.

IWA Publishing
Alliance House, 12 Caxton Street
London SW1H 0QS, United Kingdom
Tel: +44 (0) 20 7654 5500
Fax: +44 (0) 20 7654 5555
www.iwapublishing.com
publications@iwap.co.uk

Library of Congress Catalog Card Number: 2010921254
IWAP ISBN: 978-1-84339-394-8/1-84339-394-8

CH2M HILL Canada Limited

The research on which this report is based was developed, in part, by the United States Environmental Protection Agency (EPA) through Cooperative Agreement No. CR-83155901-1 with the Water Environment Research Foundation (WERF). However, the views expressed in this document are not necessarily those of the EPA and EPA does not endorse any products or commercial services mentioned in this publication. This report is a publication of WERF, not EPA. Funds awarded under the Cooperative Agreement cited above were not used for editorial services, reproduction, printing, or distribution.

ACKNOWLEDGMENTS

The Water Environment Research Foundation (WERF) thanks the volunteer efforts of the Project Subcommittee for the support they gave to the preparation and review of this report. Without their generous contribution of time and expertise, this project would not have been possible. Research for this report was conducted under WERF contract 05-CTS-3.

Research Team

Principal Investigator:
Julian Sandino
Drury Whitlock
CH2M HILL

Project Team:
John Novak, *Virginia Polytechnic Institute and State University*
Tania Datta, *CH2M HILL*
Bruce Johnson, *CH2M HILL*
Li Lei, *CH2M HILL*

Project Subcommittee

John Barber, *PSC Chair, Eastman Chemical Company*
Tony Lau, *Infilco Degremont, Inc.*
Steve Rogowski, *Metro Wastewater Reclamation District of Denver*
David Stensel, *University of Washington*
Ron Watt, *Dow Chemical Company*

Special Appreciation for Contributions

Marc Roehl, *Siemens AG*
Keith Panter, *Simon-Hartley Cambi UK*
Rob Stephenson, *Paradigm Environmental Technologies Inc.*
Royce Hammitt, *City of Des Moines*

Water Environment Research Foundation Staff

Director of Research: Daniel M. Woltering, Ph.D.
Program Director: Lauren Fillmore, M.S.

ABSTRACT AND BENEFITS

Abstract:

The process innovations evaluated in this research may reduce or eliminate the waste-activated sludge (WAS) fraction of wastewater treatment residuals, which is especially relevant for industrial facilities and smaller municipal plants which frequently have no primary sludge, but generate large quantities of WAS that are difficult to stabilize and dewater.

This project builds on the technology review previously conducted under WERF's, *Evaluation of Feasibility of Methods to Minimize Biomass Production from Biotreatment* (00CTS10T, Stensel and Strand, 2004). In this study, researchers developed a broadly applicable approach to evaluate the rapidly developing technologies in this field. This consistent approach is needed to help better understand the issues that must be considered when evaluating these systems. These factors include not only the economic, but also the noneconomic and the theoretical basis of the technologies.

Benefits:

- Provides insight into research and emerging technologies associated with WAS reduction. Data analysis from several technologies indicates positive results. One of the most interesting aspects of the study is that performance of the same technology is significantly different at separate facilities.
- Documents the state-of-the-art and full-scale performance by some of the technologies.
- Improves understanding of why some sludges are more susceptible to the technologies than other sludges.
- Provides operators of treatment facilities with information that can be used to optimize systems by producing higher-quality effluent and better-quality sludge; increase in-plant energy production; reduce sludge disposal and reuse requirements; and decrease life-cycle costs.
- Provides an initial step in a holistic analysis of the fundamental work that equipment manufacturers, design specialists, and utilities are undertaking to move these technologies forward.

Keywords: Waste-activated sludge reduction, biosolids management, anaerobic digestion preconditioning, volatile solids reduction, wastewater treatment residuals management.

TABLE OF CONTENTS

LIST OF TABLES

LIST OF FIGURES

LIST OF ACRONYMS

μm	Micrometer(s)
ABR	Anaerobic Baffled Reactors
ADM1	Anaerobic Digestion Model No. 1
ASM	Activated Sludge Model
ATAD	Auto Thermal Aerobic Digestion
ATP	Adenosine Triphosphate
BOD	Biochemical Oxygen Demand
BOD5	5-Day Biochemical Oxygen Demand
°C	Degree(s) Celsius
CBOD	Carbonaceous Biochemical Oxygen Demand
CBOD5	Carbonaceous Biochemical Oxygen Demand 5-Day Test
CEPT	Chemically Enhanced Primary Treatment
CFR	Code of Federal Regulations
CH4	Methane
CHP	Combined Heat And Power
CNP	4-chloro-2-nitrophenol
CO2	Carbon Dioxide
COD	Chemical Oxygen Demand
CODr	Chemical Oxygen Demand Reduction
dB	Decibel
DCWASA	D.C. Water and Sewer Authority
DNA	Deoxyribonucleic Acid
DNP	2,4 dinitrophenol
DS	Dewatered Sludge
EBPR	Enhanced Biological Phosphorous Removal
EEH	Enhanced Enzymatic Hydrolysis
EH	Enzymatic Hydrolysis
EPS	Exopolymeric Substances
ESP	Extracellular Polymeric Substances
Fe	Iron

FP	Focused Pulse
ft	Cubic Feet
g	Gram(s)
g VSS/g	Gram(s) Volatile Suspended Solids Per Gram
g/L	Gram(s) Per Liter
GHG	Greenhouse Gas
H2O	Water
HRT	Hydraulic Retention Time
IR	Interchange Reactor
IWA	International Water Association
JWPCP	Joint Water Pollution Control Plant
^{o}K	Degree(s) Kelvin
Kdis	Disintegration Rates
kg	Kilogram(s)
Khyd	Hydrolysis Rate
kHz	Kilohertz
kPa	Kilopascal(s)
kV	Kilovolt(s)
kW	Kilowatt(s)
kWh	Kilowatt-Hour(s)
L/h	Liter(s) Per Hour
L/s	Liter(s) Per Second
lb	Pound(s)
m^3	Cubic Meter(s)
m^3/d	Cubic Meter(s) Per Day
MBBR	Moving Bed Bioreactor
MBR	Membrane Bioreactor Process
MCA	Multi-Criteria Analysis
MCAB	Membrane-Coupled Anaerobic Bioreactor
MD	Maleic Dehydrate
MG	Million Gallons
mg/L	Milligram(s) Per Liter
MGD	Million Gallons Per Day

MHz	Megahertz
ML VSS	Mixed-Liquor Volatile Suspended Solids
mL/g	milliliter(s) per gallon
MLD	million liter(s) per day
MLE	Modified Ludzack-Ettinger
MLSS	Mixed-Liquor Suspended Solids
mm	Millimeter
MSABP™	Multi-Stage Activated Biological Process™
mV	Millivolt(s)
N2	Nitrogen
NaOH	Sodium Hydroxide
NH4-N	Ammonium Nitrogen
NOx	Nitrogen Oxide
NWWRP	Northwest Water Reclamation Plant
O&M	Operation And Maintenance
O3	Ozone
OA	Oxic-Anoxic
OCSD	Orange County Sanitation District
ORP	Oxidation-Reduction Potential
OSA	Oxic-Settled-Anoxic
OUR	Oxygen Uptake Rate
PAO	Phosphorus-Accumulating Organisms
PET	Polyethylene Terephthalate
pNP	Para-Nitrophenol
POTW	Publicly Owned Treatment Works
PS	Primary Sludge
psig	Pound(s) Per Square Inch Gauge
PVA	Polyvinyl Alcohol
PWI	Premier Wastewater International
RAS	Return-Activated Sludge
RDP	RDP Technologies Inc.
ROPEC	Robert O. Pickard Environmental Centre
rpm	Revolution(s) Per Minute

SBR	Sequencing Batch Reactor
sCOD	Soluble Chemical Oxygen Demand
SF	Soluble Fermentable Substrate
SFPUC	San Francisco Public Utilities Commission
SRT	Solids Retention Time
Ss	Readily Biodegradable COD
SUI	Scientific Utilization Inc.
SVI	Sludge Volume Index
tCOD	Total Chemical Oxygen Demand
TCP	2,4,5 trichlorophenol
THP	Thermal Hydrolysis Process
TKN	Total Kjehldahl Nitrogen
TLW	Trucked Liquid Waste
TN	Total Nitrogen
TOC	Total Organic Carbon
TOX	Kuraray Total Oxidation
TP	Total Phosphorus
TS	Total Solids
TSS	Total Suspended Solids
TVOSC	Total Volatile Organic Sulfur Compounds
TWAS	Thickened Waste Activated Sludge
UASB	Upflow Anaerobic Sludge Blankets
U.K.	United Kingdom
U.S.	United States
U.S. EPA	United States Environmental Protection Agency
VFA	Volatile Fatty Acid
VSr	Volatile Solids Reduction
VSS	Volatile Suspended Solids
WAS	Waste-Activated Sludge
WERF	Water Environment Research Foundation
WRF	Wastewater Reclamation Facility
WWTP	Wastewater Treatment Plant
Xch	Carbohydrates

Xdaa	Anoxic/Aerobic Biomass
Xi	Non-Biodegradable Particulate Chemical Oxygen Demand
Xli	Lipids
XPHA	Phosphorus-Accumulating Organisms Storage Product
Xpr	Proteins
Xs	Slowly Biodegradable Particulate Chemical Oxygen Demand

EXECUTIVE SUMMARY

Introduction

Over the previous decade, several technologies and process innovations have been proposed to reduce or eliminate the waste-activated sludge (WAS) fraction from the treatment process. Although some of these processes have been effective at one or more locations, the specific wastewater or sludge characteristics that make these effective processes are for the most part unknown. Presently, the technology is either installed without knowledge of how successful it will be or extensive piloting is needed to determine its applicability. In this study, researchers developed a broadly applicable approach to evaluate the rapidly developing technologies in the field of sludge reduction. A consistent evaluation approach helps the sector better understand the issues, not only economic, but also the noneconomic and the theoretical basis of the technologies that must be considered when evaluating these systems

This project builds on the technology review conducted under WERF's previous research, *Evaluation of Feasibility of Methods to Minimize Biomass Production from Biotreatment* (00CTS10T, Stensel and Strand, 2004). The research focuses on:

- WAS-reduction technologies that are implemented within the liquid treatment bioreactor.
- Technologies that improve the digestion of WAS in downstream aerobic and anaerobic digestion processes.

Researchers selected specific technologies to study in depth under this project based on their ability to represent the basic underlying mechanistic principles of these technologies, the availability of both field and laboratory data for these processes, the participation of vendors or utilities in providing data and samples, and the knowledge of these processes by team members.

Project Goals

This project dealt with a number of aspects of WAS-only–reduction technologies for both industrial and municipal wastewater treatment applications. The objectives of this project include:

- Develop an evaluation methodology that can be used to independently to assess the effectiveness of WAS-reduction technologies.
- Demonstrate this methodology with at least one WAS-reduction technology.

This study had primary goals of establishing the degree of WAS reduction and corresponding capital and operation and maintenance (O&M) costs, but also included details such as impacts on dewaterability (e.g., changes in polymer requirements and maximum solids content achievable), changes in volatile solids reduction and corresponding biogas production in anaerobic digestion, possible odor issues in terms of in-plant processing requirements or ultimate product quality for disposal that result from these processes, and the change in characteristics of the recycle streams back to the main process (such as increased nutrient return, increased total suspended solids [TSS] return, phosphorus removal, etc.). In addition to technical parameters, the adopted approach also considered evaluations of operability, reliability, and maintainability on each leading processes. Some of these effects were determined by laboratory testing and plant data evaluation. Others were investigated through comprehensive modeling using standard industry models such as Activated Sludge Model (ASM) 2d for liquid-stream biological treatment and the Anaerobic Digestion Model No. 1 (ADM1).

A key objective is an impartial validation of these technologies and the development of a methodology for assessing additional technologies that currently do not exist but could be developed in the future. This requires not only real-world operating data but also a degree of understanding of the fundamental mechanism behind the process. As such, a critical part of this project involves the discussion of the potential underlying mechanisms for each of the validated technologies.

Literature Review

The literature review summarizes the processes reported in the literature, and gives the status of development, knowledge of the mechanisms of sludge reduction, range of installations or testing, and ongoing developments. The review was a basis used to develop the evaluation methodology for this project so that considerations for technologies that may not currently be commercially viable but might eventually be considered for full-scale applications can be adequately evaluated. It also provides an up-to-date bibliographic reference list as a valuable resource for those interested in this subject matter.

The state of development of sludge minimization technologies continues to advance, both for applications within the liquid treatment process and for digester pretreatment. Most technologies have full-scale applications on industrial and municipal wastewater in Europe and Australasia, with a much slower implementation rate in North America. The exceptions to this trend are the combination biological processes, such as Cannibal®, which have been developed in North America and found wider acceptance. However, many technologies have a limited number of installations (e.g., chemical treatment technologies), and some technologies have shown mixed performance, particularly on applications in North America (e.g., physical pretreatment for digestion). Sludge-reduction performance reported in the literature ranges from around 20% for some of the physical lysis processes to 80% for combination biological processes and some chemical treatment processes.

Selection of Technologies to Represent General Mechanistic Principles

One of the main objectives of the literature review was to identify a limited number of technologies that represent the key mechanistic principles (biological, chemical, and mechanical) that categorize most of WAS reduction technologies for further consideration in the following tasks. The criteria for selection of representative technologies are:

- Representative of other technologies within category
- Number of operating/full-scale installations
- Length of time technology has been in operation
- Availability of data
- Technology suitable for industrial wastewater
- Technology suitable for municipal wastewater
- Ease of retrofit

Researchers obtained vendor information and conducted independent verification. Although not all of the individuals were available for interview, an attempt was made to contact the representative installations to obtain high-level information in terms of process operation and performance. From the information gathered and the evaluation criteria developed, the research team developed a short list of technologies and processes. Based on the team's experience of the technology and the information obtained from the vendors and operations personnel at the representative facilities, the team integrated the technical criteria into a matrix and rated them. In addition, the team developed criteria for the multi-criteria analysis (MCA), which is an initial step in the ultimate technology evaluation tool.

The following lists presents the technologies that were chosen, the mechanistic principle category they represent, and the associated vendor. These technologies also provided operating plant information as well as sludge samples that were used for the laboratory testing program conducted as part of this project by Dr. John Novak from the in Virginia Polytechnic Institute.

- Biological: Combination process – Cannibal® process (Siemens AG)
- Physical: Thermal hydrolysis – Cambi® process (Cambi AS)
- Physical: Pressure release – Crown® Disintegrator (Biogest AG)
- Physical/Chemical: Chemical conditioning and homogenization – MicroSludge® (Paradigm Environmental Technologies, Inc.)

The research team also obtained and evaluated data from a Microsludge® demonstration study conducted at the Des Moines, Iowa, Water Reclamation Facility (WRF).

Laboratory Scale Evaluations

Researchers carried out pretreatment and digestion studies to evaluate the effects of pretreatment on the rate and extent of solids reduction for the selected representative technologies. In addition, the team measured soluble chemical oxygen demand (COD), and for one of the sludges, they obtained electron micrographs following treatment. Both the solubilization of COD (and biochemical oxygen demand [BOD]) and the destruction of activated sludge flocs are indicators of the physical performance of pretreatment methods; however, these indicators alone are not indicative of improved digestion or solids reduction. Therefore, a goal of the project was to evaluate pretreatment technologies and compare the results to the digestion rate and extent of solids destruction.

Results from the laboratory studies indicate that some of the methods used typically in the sector to indicate the efficacy of pretreatment methods – specifically methods to increase soluble COD and destroy floc structures – cannot be used as indicators that pretreatment will increase solids destruction. Further, it appears that many of these pretreatment methods, when used at low to moderate application doses, will not improve degradation. Therefore, a complete assessment of the benefits of pretreatment technologies for a specific treatment plant appear to require pilot-scale or extensive laboratory studies.

Waste Solids Technology Evaluation Tools

Researchers designed a suite of tools as part of the WAS-reduction technology evaluation methodology which includes a facility sampling and laboratory test protocol to determine relevant sludge characteristics; a process simulation approach to facilitate the simulation for a given facility; and an MCA and life-cycle cost module for decision making that allows the consideration of technical, financial, and nonfinancial criteria.

Facility Sampling and Sludge Testing Protocol

Researchers developed a testing protocol to support the set of sludge characteristics recommended to assess the impact of sludge-reduction processes for a particular wastewater treatment facility. The team collected sludge and wastewater samples from several installations as part of this study. Along with the understanding of the basic mechanistic principles behind each of the alternatives, this served as the basis for the recommended protocol to assess relevant sludge characteristics for any treatment facility. The protocol is an appendix to this report.

Process Simulation Approach

The project discusses a general framework to simulate WAS-reduction technologies and presents a recommended general process modeling approach for them. The proposed process approach relies on the use of nonproprietary models from the International Water Association (IWA): ASM 2d for liquid stream biological process and ADM1 for anaerobic digestion. This approach must be calibrated to site-specific conditions to assist in better understanding of a particular sludge minimization technology, evaluating its applicability for a particular situation, and establishing important process parameters.

Multi-Criteria Analysis and Life-Cycle Cost Module

Capital acquisition projects (such as the incorporation of a WAS-reduction technology to an existing wastewater treatment facility) present a wide array of challenges to utility managers. Without resource constraints, capital spending priorities would be straightforward. However, in reality, utility managers face allocating limited resources across a wide range of potential projects. This allocation process is often complicated by competing policy objectives. To move forward from this situation of too many projects and inadequate resources, a cost-effective capital projects prioritization tool is needed.

This project presents a recommended approach that can be used to conduct an MCA and corresponding life-cycle cost analysis of WAS-reduction technologies. The module has been developed integrating sustainability principles. It is becoming more widely accepted to use sustainability methodologies as part of planning and decision making. The objective of the module is to enable individuals to perform a triple-bottom-line (i.e., economic, environmental, and social) analysis of the variables involving different WAS-reduction technologies under consideration. By using the tenets of sustainability for criteria development, a more comprehensive system is available to make decisions and document those decisions for public and stakeholder information.

The MCA and Life-Cycle Evaluation Tool is configured as a Microsoft Excel spreadsheet. This tool allows the user to conduct cost-effective capital project prioritization for as many as 50 different projects scored by as many as 30 different criteria. The tool uses the MCA to compare benefits, cost, and benefit-cost for user-defined WAS-reduction technology alternatives and presents rankings of benefits and benefit-costs in tabular and chart form.

Discussion and Conclusions

Sludge reduction mechanisms generally fall into the following three categories: 1) mechanisms that solubilize sludge solids and lyse cells, thereby increasing the *rate* of degradation; 2) mechanisms that render the nondegradable organic fraction degradable, thereby increasing the *extent* of degradation; and 3) mechanisms that result in the generation of less sludge by process modification. For the first two categories, most sludge-reduction vendors claim to do both. However, it is not clear that this is accomplished based on the data available from this study. Data from other investigations have found an increase in both the rate and the extent of degradation. It is also not clear to what extent cell lysis accounts for increases in solids reduction. For the third category, modification in the activated sludge process, usually performed on the return sludge recycle stream, can directly reduce solids (e.g., by chemical oxidation) or can solubilize solids in the same manner as Mechanisms 1 and 2, resulting in rapid degradation when the recycle stream reenters the aeration basin.

This research was undertaken for several reasons. The first reason was to determine whether there were indicators either in the wastewater constituents, in the biological sludge characteristics, or in the plant operation that would provide insight into the potential for solids-reduction technologies to be successful. Although it was thought that some insight into the relationship between the degree of biological solids reduction and wastewater and sludge characteristics was possible, this was considered a "long shot," primarily because the ability to obtain enough data to make comparisons across a number of plants using solids reduction technology was limited. Few solids reduction processes are in operation, and even fewer are willing to share data or samples.

The second reason for the study was to investigate mechanisms for solids reduction. This was conducted using data from both full-scale facilities as well as laboratory units where sludge reduction technologies could be compared with samples that were not treated or processed.

The third goal of the study was to develop a modeling approach that could predict solids reduction if adequate data were provided. This would include distinguishing between processes that increased the rate of degradation and those that increased the extent of degradation. The model would also include processes that increased cell lysis and those that primarily solubilized particulate organic matter. To that end, this project has developed a general framework for simulating WAS-reduction technologies using generally accepted and used models.

Finally, a fourth goal of this research, developed once the project was underway, was to determine the studies that would be needed to fully evaluate sludge-reduction technologies in a manner that would provide results independent of vendors claims and would be universally applicable to the range of sludges encountered in practice.

The state of development of sludge minimization technologies continues to advance, both for applications within the liquid treatment process and for digester pretreatment. Most technologies have full-scale applications on industrial or municipal wastewater in Europe and Australasia, with a much slower implementation rate in North America. The exceptions to this trend are the combination biological processes such as Cannibal®, which has been developed in North America and has found wider acceptance. However, many technologies have a limited number of installations (e.g., chemical treatment technologies), and some technologies have shown mixed performance, particularly on applications in North America (e.g., physical pretreatment for digestion).

This study provides valuable insight into cutting-edge research and emerging technologies associated with WAS minimization. Data analysis from several technologies indicates positive results although the performance of the same technology is often significantly different at separate facilities. Although there is still much to understand, this study improves the understanding of why some sludges are more susceptible to these technologies than other sludges. The success of a technology depends understanding the mechanisms and process parameters; particular waste stream characteristics and economic conditions; the use of appropriate operational parameters, such as adequate input energy for physical processes; and the economics of heat, sludge disposal, existing infrastructure, equipment, and chemical costs. As the technologies evolve and more information and data are gathered, there will be a better understanding of the efficacy and the underlying mechanisms. This project is an essential step in the overall research program to better understand WAS-reduction technologies. The information gathered through this investigation will help form the basis of that future research.

CHAPTER 1.0

INTRODUCTION

Biological treatment has proven to be the most cost-effective approach to reducing organic material and nutrients from municipal and industrial wastewater streams. Typically, aerobic processes oxidize the biodegradable organic material, resulting in by-products such as carbon dioxide (CO_2) and water and provide for growth of the biomass. To produce a clean effluent, most of the solids are removed, including excess biomass, inert inorganic material, and non-biodegradable volatile solids. The handling and disposal of this material (typically referred to as sludge) is an increasingly costly portion of the operation of a wastewater treatment plant (WWTP) and may be an even greater future challenge given the trend of decreasing availability and increasing costs of ultimate use and disposal options. Over the previous decade, several technologies and process innovations have been proposed to reduce or eliminate the waste-activated sludge (WAS) fraction of these residuals from the treatment process. This is especially relevant for industrial facilities and smaller municipal plants that frequently have no primary sludge (PS) but generate large quantities of difficult-to-process waste biomass. Some of these activated sludge-reduction processes and technologies are proven only in the laboratory or at the pilot scale, while others have been installed at operating treatment plants. The following are needed for most of these technologies:

- A better understanding of the fundamental mechanisms behind these technologies.
- A critical evaluation of their performance.
- A method to evaluate current and future technologies or processes for their technical and economic applicability to specific wastewaters and local economic and noneconomic conditions.

Although some of these processes have been shown to be effective at one or more locations, the sludge characteristics that make these more or less effective are for the most part unknown. At present, the technology is installed without knowledge of either how successful it will be or whether extensive piloting is needed to determine its applicability. What is needed is an approach that considers the wastewater and sludge characteristics and local economic factors of a specific treatment plant without requiring extensive pilot testing to determine whether the process will be successful and cost effective under local conditions.

Building on the technology review previously conducted under WERF's, *Evaluation of Feasibility of Methods to Minimize Biomass Production from Biotreatment* (00-CTS-10T, Stensel and Strand, 2004), an approach to help the industry better understand the mechanisms when evaluating these systems is needed. The factors include economic as well as the theoretical basis of the technologies. The general classes of WAS-reduction technologies falls into the following general categories:

- WAS reduction from the liquids treatment bioreactor

♦ Digestion pretreatment technologies
♦ Digestion technologies

Figure 1-1 illustrates where different WAS-reduction technologies are typically located in a wastewater process train. Digestion technologies were eliminated from detailed consideration during this study. In addition, dewatering technologies, such as centrifuges or belt filter presses, have not had a high degree of recent innovation, so the most useful information could be developed by considering the effects of WAS-reduction technologies on the dewatering of the resultant biosolids.

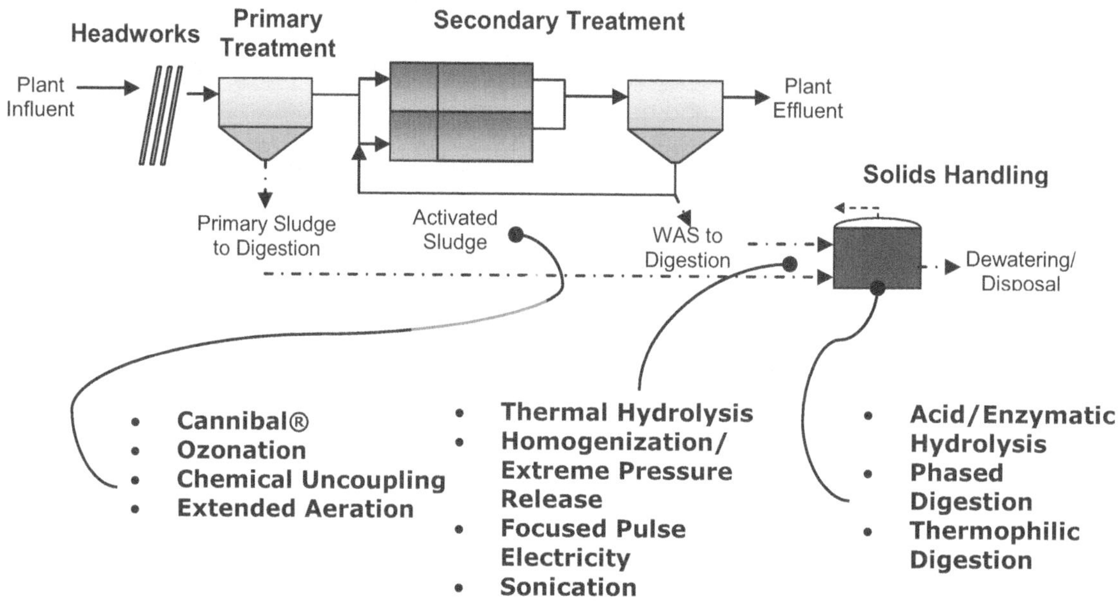

Figure 1-1. Examples of Waste Solids Reduction Technologies.

The focus of this research will be on the following:

♦ WAS-reduction technologies that are implemented within the liquid treatment bioreactor
♦ Those technologies that improve the digestion of WAS in downstream aerobic and anaerobic digestion processes

The research team selected three specific technologies to study in depth in this project based on each technology's ability to represent the basic underlying mechanistic principles, the availability of both field and laboratory data for these processes, the participation of vendors or utilities in providing data and samples, and the knowledge of these processes by team members.

CHAPTER 2.0

LITERATURE REVIEW

2.1 Literature Review Objectives

The main objectives of this literature review are to update the extensive similar effort conducted under WERF's 00-CTS-10T project, and to identify and prepare a list of the technologies that have at least one full-scale or substantial pilot installation. This literature review places into perspective the basic mechanisms behind the different WAS-reduction technologies (i.e., biological, chemical, and physical) finding full-scale applicability worldwide and the location for their application within the treatment facility (i.e., digestion pretreatment, treatment of activated sludge recycle streams). This review provides a summary of the processes reported in the literature, status of development, knowledge of the mechanisms of sludge reduction, range of installations or testing, and ongoing developments. The review was intended to develop the evaluation methodology for this project so that considerations for technologies that may not currently be commercially viable but might eventually be considered for full-scale applications can be adequately evaluated.

2.2 Context of WERF Project 00-CTS-10T

WERF report 00-CTS-10T developed an in-depth literature review of technologies to reduce biomass production in biological liquid treatment processes. Following the literature review, bench-scale tests were conducted to confirm whether including an anaerobic zone in the activated sludge process for enhanced biological phosphorous removal (EBPR) provided a reduction in WAS generation. An economic evaluation of ozonation of return-activated sludge (RAS) was also conducted.

The report by Stensel and Strand (2004) was structured differently than the current literature review, as it was intended to be the final product for that project. The following eight technologies were evaluated, each with a dedicated chapter:

- Anaerobic treatment
- Sequential anaerobic/aerobic treatment
- Anaerobic/aerobic EBPR
- Cell disruption by mechanical and thermal methods
- Cell disruption by chemical methods
- Chemical uncoupling processes
- Predator enhancement processes
- Extreme solids retention time processes

However, this current literature review is only the first step in developing an evaluation methodology toolkit for assessing sludge minimization technologies; therefore, the technologies have been grouped based on whether the main mechanism for sludge reduction is biological, chemical, or physical. This literature review includes some technologies that were not included in the Stensel and Strand (2004) report, such as Cannibal® and ultrasound. It also provides updated information on industrial and municipal installations and on the mechanisms for the technologies discussed in the Stensel and Strand (2004) report. The only area where minimal progress has been made is in the chemical uncoupling technologies. Also, the EBPR process was not included in the current review as the previous report documented no measurable advantages. Thus, the results and conclusions from this literature review will therefore differ in some aspects from those provided in the previous report.

2.3 Biological Processes

2.3.1 Process Fundamentals and Mechanisms

The primary function of suspended and fixed-film biological reactors is to convert biodegradable organics to suspended solids by creating an environment where biota can use the organic material as a food source. Especially in aerobic environments, an undesired product of this reaction is the creation of excess biota via anabolic metabolism. Carbon dioxide, nitrate, nitrogen gas, and water are other examples of by-products of the cellular metabolism in an aerobic reactor. Anaerobic reactors reduce COD to methane and CO2. In an anaerobic environment, the cellular growth rate is slower, and the metabolic reactions inherently result in lower yields; therefore, less excess biota are produced. Ideally, the aerobic reaction would be mostly catabolic with minimal solids wasting.

In the activated solids process, biodegradable organic material in the influent wastewater is used by microbes, with some converted to cellular material for growth and reproduction of the microbes and some used as energy sources to maintain the microbes' metabolism, with by-products, such as CO_2 and water. In activated solids systems, the WAS observed yield is normally between 0.7 and 1.2 kilograms per kilogram (kg/kg) BOD. The treatment process therefore produces excess biomass, although the yield is often less than the value of the influent organic load. The organic material contained in the biomass after death and decay is slowly biodegradable, largely due to the time required to break down the cell walls. In mesophilic anaerobic digestion, this may take eight days or more. All types of solids-reduction technologies operate by increasing the consumption of the biomass decay products as a food source, thus reducing the overall production of solids. This consumption may take place within the activated solids process, in a downstream digester, or in an additional secondary treatment stage.

Biological sludge reduction processes have for the most part focused on the following four mechanisms:

1. Use of anaerobic wastewater treatment processes, which have a lower biomass yield than aerobic processes
2. Predation, where microbes that grow in a facultative environment break down biomass produced in an aerobic environment and vice versa
3. Sequencing anaerobic and aerobic digestion processes, as different fractions are biodegradable in each environment
4. Biological cell lysis, either through endogenous decay in aerobic environments or acid phase hydrolysis in anaerobic environments

Some processes may include more than one mechanism, providing multiple approaches for sludge reduction. Anaerobic processes, or those that include environments with low oxidation-reduction potential (ORP), will result in lower biomass production in those environments. Table 2-1 shows the typical biomass yields for different ORP environments.

Table 2-1. Typical Yields under Different Oxidation-Reduction Environments.
Reprinted with permission from Tchobanoglous et al., 2003.

Environment	Electron Acceptor	Yield (g VSS/g COD)	Yield (g VSS/g CBOD)*	Products
Aerobic	Oxygen	0.40	0.80	CO_2, H_2O
Anoxic	Nitrate	0.30	0.60	N_2, CO_2, H_2O
Anaerobic	Organics	0.06	0.12	Organic acids
	CO2	0.05	0.10	CH_4
	Total	0.11	0.22	CO_2, CH_4

Notes:
CBOD = carbonaceous biochemical oxygen demand
CH_4 = methane
CO_2 = carbon dioxide
COD = chemical oxygen demand
g VSS/g = gram(s) volatile suspended solids per gram
H_2O = water
N_2 = nitrogen
* Assumes COD:CBOD 2:1, typical of municipal wastewater. Ratios for industrial wastewater are widely variable.

2.3.2 Anaerobic Treatment of Wastewater

Anaerobic treatment is a proven process for high-strength waste streams and for concentrated streams, such as upflow anaerobic sludge blankets (UASB) for high strength industrial wastewater. Full-scale treatment of municipal wastewater has been implemented in some countries in South America and Asia where wastewater temperatures are normally above 20 degrees Celsius (°C). Many of these facilities have an aerobic polishing step to meet effluent discharge BOD and total suspended solids (TSS) concentrations (Sandino et al., 2004). Configurations include various attached growth processes, anaerobic baffled reactors (ABR) where the wastewater flows up through sludge blankets in a series of reactors, UASB, and membrane-coupled anaerobic bioreactors (MCABs). These processes aim to reduce production of biomass rather than reduce biomass after it has been produced. However, research shows that performance declines at lower temperatures and at lower organic loading rates. Bench and pilot-scale tests have been conducted with MCABs to provide better effluent quality. However, sustainability of membrane flux under these conditions is yet to be verified.

At temperatures above 20°C, full-scale municipal wastewater treatment with UASBs has shown COD removal rates of 50-80% and TSS removal rates of 50-76%. Waste solids produced may be 50 to 75% less than conventional primary and secondary treatment trains. Bench-scale tests at lower temperatures (7-18°C) have reported COD removal rates in the range of 45-85%. Longer hydraulic retention times (HRTs), up to 14 hours, are required at lower temperatures to meet reasonable removal rates (Rogalla et al., 2006). At warmer temperatures, coupled anaerobic-aerobic liquid processes have been found to be competitive with conventional treatment trains and usually do not require separate digestion of WAS from the aerobic train, as this waste stream can be diverted to the anaerobic reactor.

Horng et al. (2006) have also provided a brief summary of a full-scale application on industrial wastewater treatment of high-strength maleic dehydrate (MD). A UASB was installed ahead of an existing activated sludge process and reduced WAS production 60%.

Biological phosphorus and ammonia removal cannot be conducted within the anaerobic process. Chemical phosphorus removal and biological ammonia removal in an aerobic reactor or through a stripping process would be required to remove these nutrients. Nitrate reduction can be achieved by returning mixed liquor from the aerobic process to the end zone of the anaerobic process. One such configuration, the Ananox® process, was demonstrated in Bologna, Italy (Garuti et al., 1992), using an anaerobic baffled reactor followed by an aerobic reactor and clarifier. Solids production was 0.2 g TSS per gram (g) COD reduction (CODr), approximately half that of a conventional activated sludge process.

Additional information on different configurations and research data can be found in the 00-CTS-10T report.

2.3.2.1 ABRs as Primary Clarifier Retrofit

Pilot- and full-scale commercialization of ABR technology as a retrofit to primary clarifiers has been undertaken by Atkins, an engineering firm based in the United Kingdom (U.K.). The concept is based on installation of baffles within a rectangular or circular primary clarifier tank. Flow velocity is designed to allow sludge blankets to be maintained in each zone, providing for development of an anaerobic biomass with long solids retention time (SRT), often in the range of 30 to 60 days. As screened wastewater flows up through each zone, particulates are retained within the sludge blanket, and BOD is metabolized by the biomass. Four or five zones are typically installed. Conditions in each zone show step changes with lower pH, higher alkalinity, and changing volatile fatty acid (VFA) concentration. Some methane production will occur in the latter zones (Figure 2-1).

A full-scale retrofit was undertaken at the Dunmore WWTP, Londonderry, Northern Ireland. Pilot-scale ABR tests have also been conducted at the Orange County Sanitation District (OCSD), Fountain Valley, California (Atkins Ltd., 2005).

The OCSD uses chemically enhanced primary treatment (CEPT), with HRT in the primary basins, varying from 2.5 to 4 hours. The ABR pilot tests were conducted under different scenarios, ranging from no chemical addition to ferric chloride and polymer addition and with HRTs ranging from 3 to 4.4 hours. A summary of the data from the pilot tests is provided in Table 2-2. The ABR effluent BOD concentration was attributed to solubilization of BOD for which there was insufficient time for conversion in the ABR. Fractionation of the BOD in the ABR effluent and OCSD primary clarifiers was not compared. There were concerns that because the ABR effluent BOD was higher than from the existing primary clarifiers, retrofitting the primary clarifiers with the ABR process could increase oxygen demand in the activated sludge process and produce more WAS.

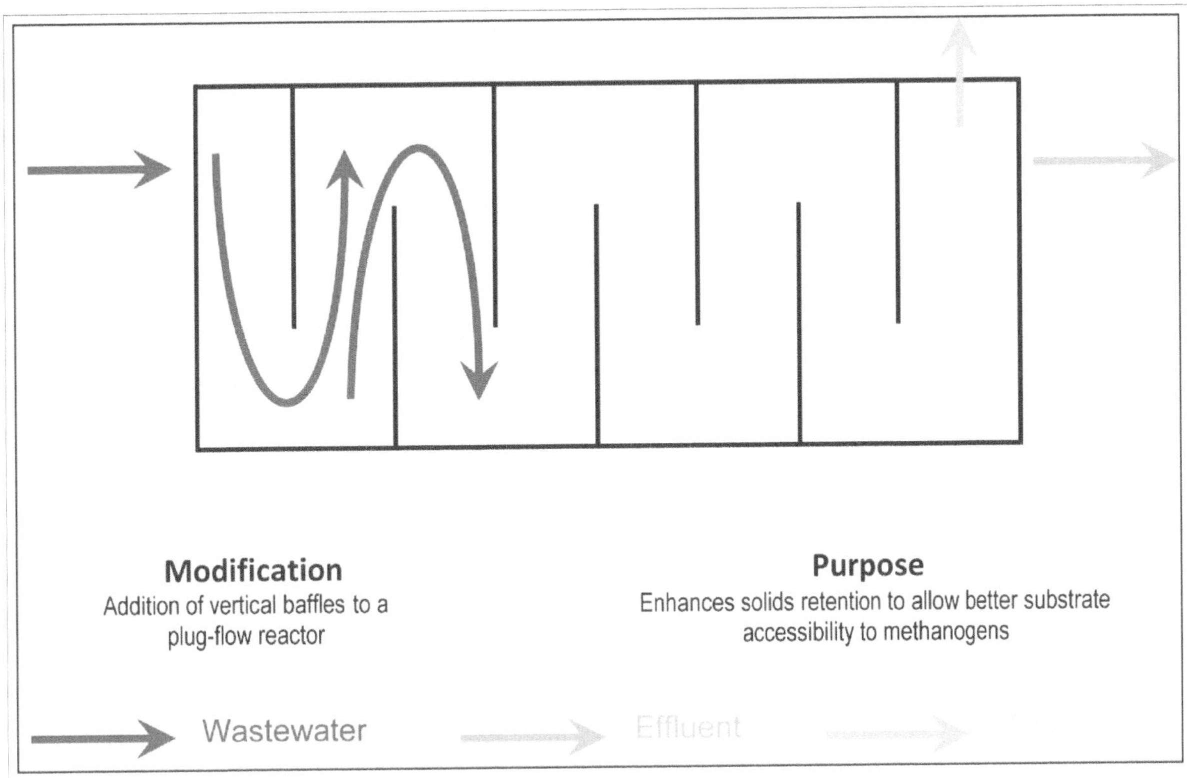

Figure 2-1. Schematic of the ABR Process.
Reprinted with permission, Atkins, 2005.

Table 2-2. Summary of ABR Pilot Plant Performance at the OCSD WWTP.

Parameter	ABR Performance[1]		OCSD Primary Clarifiers with CEPT[1]	Typical Primary Clarifiers (no CEPT)[2]
	No CEPT	With CEPT		
TSS removal (%)	52	60–65	70	50–70
BOD_5 removal (%)	25–30	25–30	45–50	25–40

NOTES:
BOD_5 = 5-day biochemical oxygen demand
CEPT = chemically enhanced primary treatment
OCSD = Orange County Sanitation District
TSS = total suspended solids
WWTP = wastewater treatment plan
[1] Atkins Ltd., 2005
[2] Tchobanoglous et al., 2003

The final phase of the project provided an evaluation of costs and potential benefits and risks for a large-scale retrofit of the existing primary clarifiers to the ABR process. As the OCSD primary clarifiers are operated at a short HRT, concerns arose about the performance of the ABR system on a long-term basis, particularly during peak flows. A sensitivity analysis showed that there may not be a cost–benefit advantage for the ABR system retrofit at a facility such as OCSD, which conducts high-rate primary clarification with chemicals (Atkins Ltd., 2007). A full-scale test was not conducted at OCSD; therefore, data do not exist to show the impacts of coupled ABR and activated sludge treatment and whether a sequenced anaerobic-aerobic treatment train would provide any additional advantages, as discussed in the following section.

2.3.3 Anaerobic-Aerobic Sequenced Treatment

2.3.3.1 Sequenced Digestion

Sequential anaerobic-aerobic digestion has demonstrated sludge reduction at bench and pilot scale. This process was investigated at laboratory scale and is known to have volatile solids removal as high as 60% and above. Moreover, significant total nitrogen (TN) and total phosphorus (TP) removal in digested solids has been reported (Kumar et al., 2006a), which implies decreased nutrient load on WWTP from dewatering return streams.

It was found during laboratory investigations that various fractions of polysaccharides and proteins are present in wastewater and WAS. These fractions and associated cations play a central role in solid digestion process (Park et al., 2006). Higgins and Novak (1997a), Novak et al. (2003), and Novak and Park (2004) found that the major cations associated with the sludge floc are sodium, potassium, calcium, magnesium, iron, and aluminum. Frølund et al. (1996) proposed a sludge floc model in which a floc consists of micro-consortia embedded in matrix of biopolymers including proteins, polysaccharide, deoxyribonucleic acid (DNA), and humic acids. Extracellular polymers prefer to bond with divalent ions as they form more stable complexes (Rudd et al., 1984). These biopolymers are bound in the sludge matrix and are associated with different cations. During aerobic digestion, divalent cations accumulate in the solution resulting from the release of lectin-like proteins, which are then degraded, while during anaerobic digestion, ferric-bound biopolymer is released upon the reduction of iron (Fe) (III) to Fe (II) and is degraded (Park et al., 2006). However, it has also been shown that the breakdown of lipids to long-chain fatty acids can also inhibit anaerobic digestion (Jeganathan et al., 2006).

Sequential anaerobic-aerobic digestion helps in removing all of the previously mentioned fractions of biopolymers and hence achieves more solid reduction than conventional "only anaerobic" or "only aerobic' digestion process. Bench tests were conducted with PS and WAS from the Blue Plains WWTP, Washington, D.C (Kumar, Murthy, and Novak, 2006b). Anaerobic digestion was conducted at mesophilic and thermophilic temperatures, followed by aerobic digestion at three, six, and nine days. Aerobic digestion at three days provided an additional 10% volatile solids reduction (VSr), and 12% was achieved at six days. Ammonia was reduced 80 to 90%, and dewaterability was improved. Endocrine-disrupting activity in the filtrate was also found to be significantly lower after the aerobic digestion step. Additionally, Kumar et al. (2006b) have shown that sludges treated under sequential anaerobic-aerobic digestion process are less odorous than conventionally treated sludges.

2.3.3.2 Sequenced Wastewater Treatment

Similar mechanisms have been shown to operate in wastewater treatment systems that use sequenced aerobic and anaerobic bioreactors (Novak et al., 2006a; and Novak et al., 2006b). Sequenced wastewater systems typically have a recycle system from the main aerobic reactor, through a side-stream anaerobic or anoxic reactor, and may be operated as oxic-anoxic (OA) systems or oxic-settled-anoxic (OSA) systems.

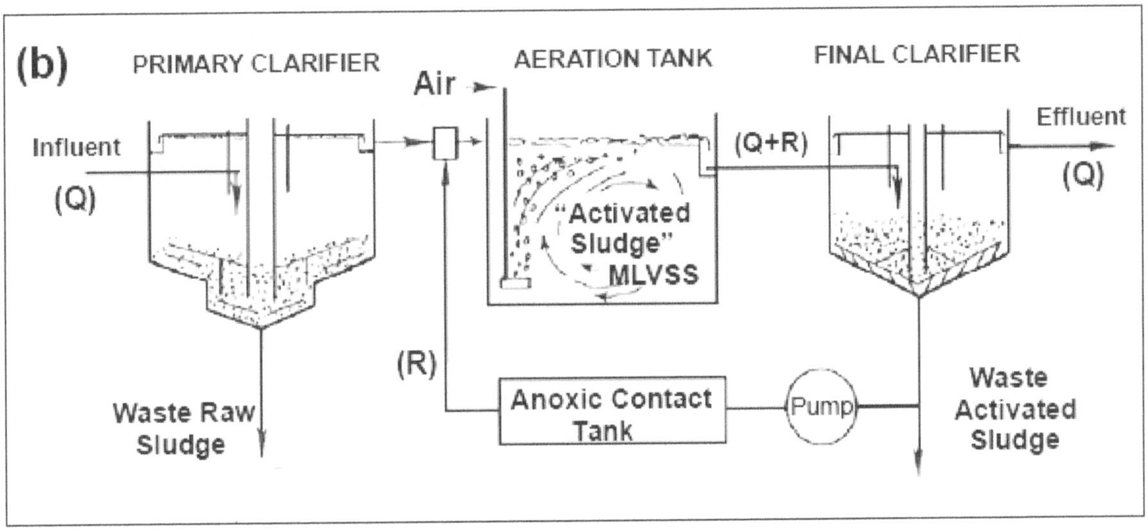

Figure 2-2. Schematic of the Sequencing Wastewater Treatment.
Reprinted with permission from the Water Environment Federation, 2004.

These systems have the added advantages of continual cycling of organic matter from one environment to another and longer SRTs than conventional activated sludge treatment systems. Research has shown that longer SRTs reduce system volatile suspended solids (VSS) yield (Moon et al., 2004). It has also been suggested that cycling between aerobic and anoxic environments results in an uncoupling between anabolic and catabolic metabolism stimulated by stress and starvation in the changing environments (Moon et al., 2004). Reported yields from sequencing systems range from 15-90% lower than aerobic systems. Some systems compared sequenced systems with aerobic systems at different SRTs. Some data are generated from systems that were only fed soluble substrates, and therefore the data would not be representative of municipal wastewater systems, which include nondegradable inert material. Data from bench-scale tests that have attempted to maintain similar SRTs in aerobic and sequenced systems have shown that for the same SRT, sequenced systems solids yields may be 10-50% less than aerobic systems.

2.3.4 Predation Processes

Predation systems are based on use of protozoa (such as amoebae and stalked ciliates) and metazoan species to consume bacteria grown in the aerobic systems. Protozoa break down bacterial cells and hydrolyze polymeric mucoproteins and cell protoplasm. These predators require a longer SRT, which may be provided through a fixed-film or dispersed-floc bioreactor (or membrane bioreactor) with biomass recycle. This approach has been pioneered in Europe at Lund University, Sweden, and demonstrated by Anox AB. The system was designed with a high-rate aerobic bioreactor with no recycle, where the HRT was close to the washout rate of the bacteria, followed by a reactor with a recycle and longer SRT to favor growth of the predators. A large-scale pilot test by Christensson and Welander (2001) showed that with municipal wastewater, the high-rate bioreactor performance was low, with COD removal of 30%. Overall yield from the predator system was 0.3 g TSS/g COD, compared with 0.35 g TSS/g COD in the conventional activated sludge process. A full-scale plant was installed on pulp and paper wastewater, with a high COD concentration of 5,200 milligrams per liter (mg/L), replacing an existing activated sludge system. Solids yield from the predation process was 0.02 to 0.05 g TSS/g COD compared with 0.14 to 0.22 g TSS/g COD previously. Research on predation systems has shown that the first-phase bacterial bioreactor HRT should be less than six hours to encourage growth of a dispersed biomass rather than filaments or dense flocs. Predation stage SRTs have ranged from 10 to 122 days, and yields have been reported to range from 20-90% less than conventional treatment.

Additional development of predation processes has been conducted by the University of Cambridge, U.K.; Lapinski and Tunnacliffe (2003) have shown that predators, in particular rotifers, can be encouraged by providing particles in the optimum size range (1 to 3 micrometers [µm]) for consumption and can reduce biomass by 10-30% and improve gravity clarification by 10%. The process has been patented (Tunnacliffe and Lapinski, 2006; United States [U.S.] Patent 7,153430 B2) and is being commercialized by Cambridge Enterprise Ltd., the technology commercialization arm of the University of Cambridge. The system uses a nozzle to provide a 4 to 6 bar pressure drop in a side stream of RAS, providing cavitation and reducing the particle size without excessive cell lysis. To prevent washout of the predators with the effluent, it is recommended that the predation reactor be separated as a side stream reactor with a filter system on the effluent to trap the predators in the tank. Target predator concentrations are 50,000 to 100,000 mg/L.

The Siemens' Cannibal® process suggests by name it is a predation process. However, the operation is different than that previously described, and it incorporates most of the biological sludge reduction mechanisms. It is therefore addressed separately in Section 2.2.6.

2.3.5 Biological Cell Lysis

2.3.5.1 Extreme SRT and Endogenous Decay

Extended aeration is a proven process for reducing sludge yields and has been implemented at many WWTPs worldwide. This process operates by allowing sufficient SRT in the activated sludge process to allow for natural cell death and breakdown of cell walls. Extended aeration processes can operate at SRTs of 20-25 days, and some plants have operated at significantly longer SRTs up to 60 days. High mixed-liquor suspended solids (MLSS) concentrations are required to provide these long SRTs without excessive tankage, and this can affect the stability of the secondary treatment process and impact performance of the secondary clarifiers. However, use of membranes for solids-liquid separation (as in the membrane bioreactor process [MBR]) removes the constraints typically associated with gravity settling in clarifiers and allows a higher MLSS concentration so that a longer SRT can be achieved within a smaller footprint.

The mechanism of endogenous decay in extreme SRT systems was first proposed by Gaudy et al. (1971), where cellular material released by endogenous decay can be metabolized by other microbes.

Kuraray Total Oxidation Process. The Kuraray total oxidation (TOX) process is a patented approach to achieve endogenous decay to eliminate sludge wasting. The process is based on the use of a moving bed bioreactor (MBBR) using polyvinyl alcohol (PVA) gel beads as the carrier, followed by a high-solids oxidation process (TOX). The vendor claims that the biofilm step is critical to the success of its process, as the sludge production rates are half that of a typical activated sludge plant, reducing the load on the TOX stage. The TOX stage operates at a high-dissolved oxygen concentration in the range of 6-7 mg/L, while the BOD load is low, as most of it is consumed in the biofilm step. Therefore, the TOX reactor operates in a starvation mode. Much of the work has been conducted using membranes for solid-liquid separation;

however, a pilot plant has been operated with gravity settling in clarifiers. These configurations are shown in Figure 2-3. The TOX process has operated at MLSS concentrations up to 25 grams per liter (g/L) with membrane separation and up to 18 g/L with secondary clarifiers. Literature provided by the vendor did not discuss operating SRTs or the size of the reactors. Removal or buildup of inert debris in the system was also not discussed.

Much of the work to date has been with industrial wastewaters with negligible influent suspended solids. One full-scale system has been in operation for five years with no sludge wasting, and additional pilot systems have been operated on industrial wastewaters. One pilot-scale trial was operated using municipal wastewater from a facility in Slovenia (Rouse et al., 2006; Rouse et al., 2007). This process claims total oxidation of the organic sludge as yielded by a biofilm process used for municipal wastewater treatment. The process involves two cycles – first for nitrogen removal and second for sludge reduction. Reported TN removal efficiency in the TOX process was 67% with 200% internal recycle at 20°C. Pilot plant studies and bench-scale experiments show that TOX is capable of achieving decay coefficient in range of 0.06-0.09 per day consistently, and steady-state data over a two-month period showed that no sludge removal was required other than that removed in the effluent. However, solids mass balances were not provided for periods with wastewater temperatures below 20°C.

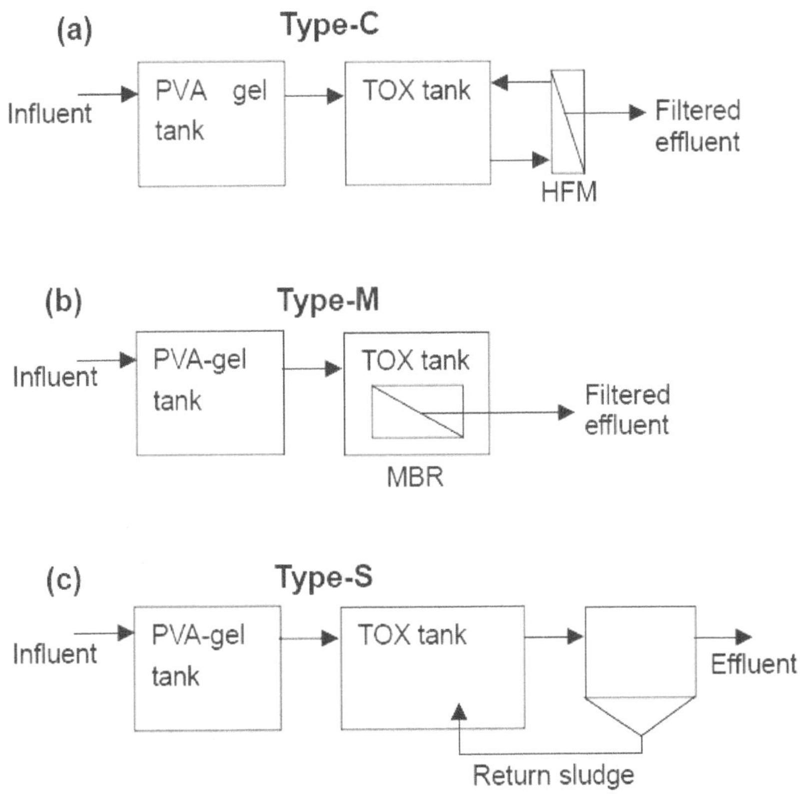

Notes: (a) Type-C (patented), hollow fiber membranes outside the sludge tank; (b) Type-M (patent pending), hollow-fiber membranes inside the sludge tank (MBR style); and (c) type-S (patent pending), standard sedimentation tank.
Source: Rouse et al., 2007

Figure 2-3. Kuraray TOX Process Based on Solids-liquid Separation Methods.

2.3.5.2 Enhanced Enzymatic Hydrolysis

The enzymatic hydrolysis (EH) process is an enhanced approach to acid/gas phase digestion, which has been developed by United Utilities in the U.K. in partnership with Monsal Ltd. The process was designed as a pseudo-plug flow approach to acid hydrolysis, with up to six acid reactors in series, each operated at 32-42°C to maximize enzyme production. To meet the U.K. enhanced treated standard[1] for pathogen destruction, the process was adjusted with the first three reactors operated at 42°C, followed by three reactors operated at 55°C to maximize enzymatic reaction rates and meet the pathogen destruction standards and was named the enhanced enzymatic hydrolysis (EEH) process (Le, 2006). A flow schematic is provided in Figure 2-4. The acid pretreatment stage is designed with a two-day HRT, followed by gas phase digestion for 10 to 14 days. The first demonstration facility was installed at the Macclesfield WWTP in 2001. Three more installations are now operational and three other facilities are being designed in the U.K. (Le, 2006). Data from three of the facilities – Macclesfield, Crewe, and Blackburn WWTPs – indicate improvements in VSr ranging from 15-31% (Le, 2007). To meet the U.S. Environmental Protection Agency (U.S. EPA) Part 503 Class A pathogen standards, the last reactors of the acid stage would need to be larger to meet the 16-hour batch holding time required at 55°C.

A further development of this technology is underway, known as the fermented cake option. The EH or EEH process can be implemented, depending on pathogen destruction regulations, with a total retention time of four to six days (Le and Harrison, 2006). The fermented sludge is dewatered. The filtrate may have VFA concentrations as high as 20,000 mg/L and can be used as a food source for nutrient removal or can be digested in high-rate UASB reactors to produce biogas for co-generation systems. Demonstration scale tests have shown that the cake is stackable, has a low odor, and meets U.K. enhanced treated standard even after 10 days' storage.

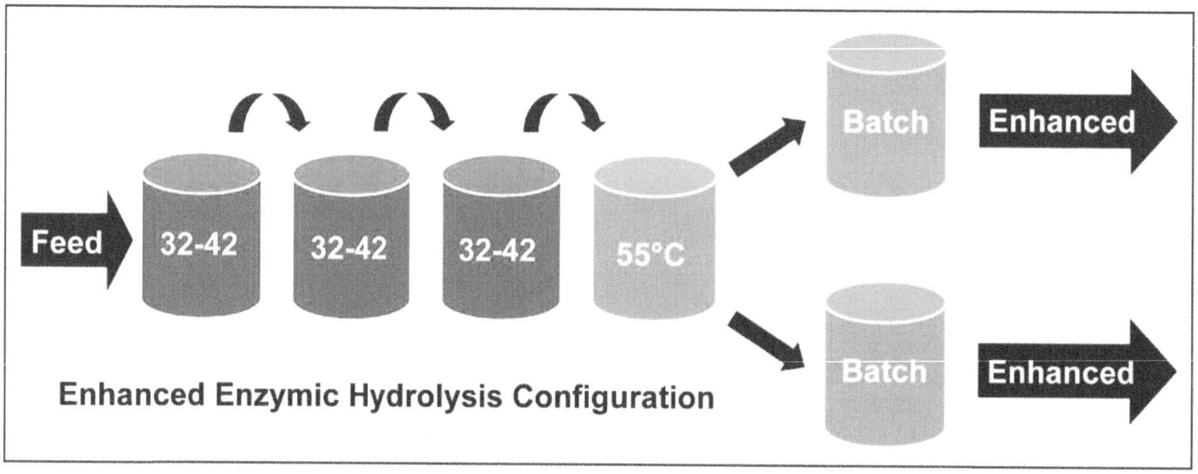

Figure 2-4. Flow Schematic of the EEH Process.
Reprinted with permission from Le, 2007.

[1] The U.K. "enhanced treated" standard requires *E. coli* reduction ≥99.9999%, <1,000 *E. coli* per g dry solid, and nondetectable *Salmonella* in 2 g dry solids.

2.3.6 Combination Processes

2.3.6.1 Cannibal® and CATABOL

The Cannibal® process is a patented process currently owned by Siemens. The process includes a reactor, known as the interchange reactor (IR), which operates in a facultative mode, on the cusp of anoxic and anaerobic environments. Rather than being wasted, the WAS is sent instead to the IR. The same volume is returned daily from the IR to the activated sludge process. Solids retention times in the system may be as high as 100 days. This process includes most of the biological sludge reduction mechanisms discussed. As the name suggests, a key mechanism for sludge reduction is based on predation, although research by Novak et al. (2006a) shows that the anaerobic-aerobic sequencing mechanisms previously described also contribute to reduced sludge volumes in the Cannibal® process. This is supported by findings reported by van Loosdrecht and Henze (1999), which summarized that it is impossible to differentiate between cell maintenance, endogenous metabolism, lysis, decay, and predation in operating systems with anoxic recycle reactors. Data from full-scale Cannibal® installations show that WAS yields below 0.1 g TSS/g BOD applied can be achieved, provided that inert solids are removed by a solids separation module. For municipal wastewater installations, the inert solids removed are typically 0.1 to 0.2 kg TSS/kg BOD (±10%)[2] applied to the bioreactor at a plant without primary clarification. Currently, at least 10 plants operate in the U.S., including nine municipal and one industrial, with capacities between 1.9 and 61 million liters per day (MLD).

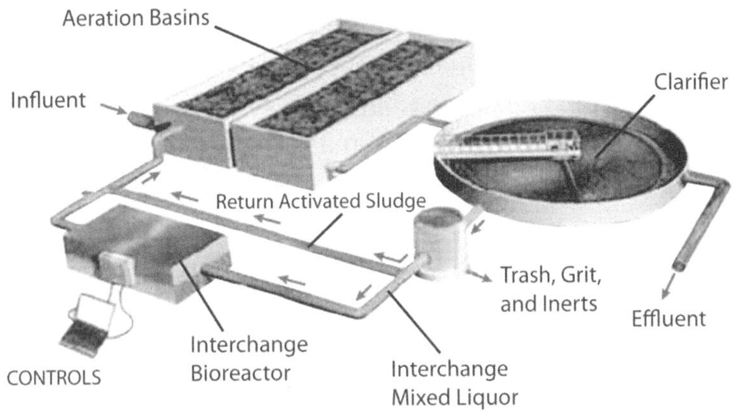

Figure 2-5. Flow Schematic of the Cannibal® Process Integration with an Activated Sludge Plant (Siemens Literature).

The Cannibal® process consists of two main processes: 1) the physical separation step and 2) the biological treatment step. Biological treatment uses an anaerobic-anoxic unit (IR) through which a portion of the recycle sludge passes prior to being returned to the aeration basin. Discussions with vendors and experts in the Cannibal® process suggest that cycling mixed liquor between the aerobic reactor and IR puts environmental stresses on microbes, which helps in killing bacteria and releases cell material for others microbes. This results in very low yield and higher nutrients removal. Through bench-scale lab test, Novak's research indicated that low yields are achievable, but various protein fractions associated with cations play very important role in the Cannibal® process (Novak et al., 2006a).

---

2 As reported by Siemens.

Physical separation using fine screens and hydrocyclones is used to remove inert material from mixed liquor in the Cannibal® process. All RAS is passed through fine screens regularly to remove inert organic material, but hydrocyclones are intermittently operated on RAS to remove heavy organic material, grit, and dense inorganic material.

Some of the early work with this type of combination process for sludge reduction was conducted by Dr. Boris Khudenko. He has developed a similar process, known as CATABOL™. Three installations, of which one is no longer operational due to industrial reuse and one pilot plant that is also not operational, have used this process. His process is reportedly the basis for the design of the 57 MLD Cartersville Plant in Georgia, which is currently under construction. Patent disputes will likely need to be resolved through the legal process.

Roxburgh et al. (2006), Mesloh et al. (2007), and Rickermann and Johnson (2007) suggest that the Cannibal® process is economically favorable for small to medium-sized plants without primary treatment and for those plants that already have aerobic digesters, which could be retrofitted for use as IR tanks. This makes this process particularly applicable to industrial wastewater treatment applications.

2.3.6.2 The Aquarius Multi-Stage Activated Biological Process (MSABP™) Process

Aquarius Technologies Inc. has developed the MSABP™, a treatment system that claims to produce no sludge. The process treats finely screened wastewater in a series of aerobic and anaerobic stages. Eight to 16 stages may be used, depending on the wastewater characteristics, each with an independently controlled air supply. A proprietary synthetic media is used to provide a fixed-film growth surface for the microbes. The process includes the principles of sequenced environments and predation to achieve sludge minimization. The process has been implemented at a few industrial and small municipal installations in Israel and the U.S. Three case studies are provided on the company Web site and are summarized in Table 2-3.

Table 2-3. Summary of Aquarius MSABPTM Case Studies.
Reprinted with permission from Aquarius Technologies, Inc.

Parameter	Case 1 – Teva Pharmaceutical	Case 2 – JH Ranch Resort	Case 3 – Kibbutz Yavne
Flow (m^3/day)	300	150	795
Influent to MSABP™			
COD (mg/L)	50,000	1,000	-
BOD_5 (mg/L)	20,000	450	350
TSS (mg/L)	2,000	250	450
TOC (mg/L)	15,000	-	-
NH_3-N (mg/L)	-	100	65
Effluent from MSABP™			
COD (mg/L)	<6,000	<60	-
BOD_5 (mg/L)	<1,000	<10	<15
TSS (mg/L)	<350	<10	<20
TOC (mg/L)	<1,500	-	
NH_3-N (mg/L)	-	<1	<1
TN	-	<10	<10

Notes:
BOD = biochemical oxygen demand
BOD_5 = biochemical oxygen demand 5-day test
COD = chemical oxygen demand
m^3 = cubic meter(s)
mg/L = milligram(s) per liter
MSABP™ = Multi-Stage Activated Biological Process
NH_3-N = ammonia nitrogen
TN = total nitrogen
TOC = total organic carbon
TSS = total suspended solids

2.3.7 Summary

A number of mechanisms exist for sludge reduction using biological processes, with varying applicability for sludge minimization. Anaerobic processes, such as UASBs, have been implemented at full scale for high-strength wastewater and in higher temperature (>20ºC) municipal applications and may achieve biomass production up to 75% lower than aerobic systems. However, anaerobic processes do not achieve as high a BOD removal as aerobic systems and typically require aerobic polishing or membrane separation to achieve direct discharge to the environment. For this reason, sequential wastewater treatment using anaerobic-aerobic sequencing has been implemented in practice and has shown up to 60% less WAS production than aerobic-only wastewater treatment. Tests with aerobic digestion following anaerobic digestion indicate that an additional 10-12% VSr may be achieved. However, much of the work on sequencing environments and impacts on biodegradability of different fractions is still at the bench-scale level.

Predation systems have been tested or implemented at a few larger-scale installations, with one installation on pulp and paper wastewater and reported waste activated sludge reduction up to 70% but more typically around 20-40%. However, the technologies developed in Europe appear to be difficult to operate in terms of selecting and maintaining sufficient concentrations of predator species and have not received much market acceptance.

Biological cell lysis through endogenous decay has historically not been widely practiced due to limitations on solids-liquid separation at high MLSS concentrations required to maintain the long SRT without excessive tankage or the incorporation of membranes. The Kuraray TOX process uses a biofilm secondary treatment process to reduce biomass production, followed by a high-concentration aerobic thermophilic digester, and chemical treatment to increase the decay rate has currently one full-scale industrial installation. The United Utilities EH process has been used as a pretreatment step for anaerobic digestion at municipal WWTPs, with four installations in the U.K. and typical increase in VSr of 25%.

The Cannibal® and CATABOL™ processes incorporate most of the biological mechanisms for sludge minimization. At least 10 plants operate in the U.S., with more in design or construction. The process has been implemented on municipal and industrial wastewaters and when properly integrated within the overall wastewater treatment process has met the target of WAS yields below 0.1 g TSS/g BOD and inert solids removal of 0.2 to 0.3 kg TSS/kg BOD at a municipal plant without primary clarification. The Aquarius MSABP™ process has a different configuration to the Cannibal® process but employs similar mechanisms. Three reported installations have occurred on industrial and municipal wastewater.

2.4 Chemical Processes

2.4.1 Process Fundamentals and Mechanisms

Chemical sludge reduction processes have focused on the following three mechanisms:

1. Cell lysis and significant cell wall weakening through extreme oxidation or pH conditions, releasing cellular organic matter for further degradation
2. Chemical uncoupling that changes the proton gradient across the cell wall, allowing continuing oxidation of organic matter but inhibiting adenosine triphosphate (ATP) generation, slowing cell growth
3. Energy spilling under stressful environments or nutrient-limited conditions

2.4.2 Chemical Lysis

2.4.2.1 Ozonation

Oxidation of RAS for WAS reduction has been studied experimentally and implemented at full scale by such companies as Praxair® and Ondeo Degremont (Biolysis® 'O'). The concept includes applying ozone to a portion of the RAS in a suspended solids-aerated biological reactor RAS stream to lyse the cells. Lysis allows biota to metabolize soluble COD (sCOD) within the conventional aerated treatment process, such as activated sludge. The increase in readily biodegradable sCOD within the aerated process does increase the oxygen uptake rate (OUR) within the system, and, therefore, aeration costs increase as a result. Theoretically, the sludge yield in these systems should be reduced to 0.36 kg TSS/kg COD removed, which is about 64% less sludge yield (Fabiyi et al., 2007). Operating data has shown measured performance ranges from 30 to 70% lower WAS production.

Studies have shown the ozone dosing rate to be in the range of 0.05 to 0.4 kg ozone (O_3)/kg total solids (TS) removed. Work by Manterola et al. (2006) indicates that at higher TSS concentrations, a lower dose of ozone will achieve a similar degree of COD solubilization as a higher ozone dose when the TSS concentration is lower. It has also been shown that gas flow rate is an important aspect of ozone mass transfer efficiency; therefore, a specific ozone dose (mg O_3/g TSS) applied at a higher gas flow rate and a lower ozone concentration will produce a higher degree of COD solubilization than the same dose at a lower gas flow rate (Manterola et al., 2006; Zhou and Smith, 2000). Bench-scale tests were conducted by Dytczak et al. (2007) to compare anoxic-aerobic sequencing batch reactors (SBRs) with aerobic SBRs. The smaller floc structure of the anoxic-aerobic system showed a higher degree of disintegration for the same ozone dose when compared with the larger, more filamentous floc structure of the aerobic system.

Table 2-4. Summary of Ozone Doses Reported in the Literature. Based on Fabiyi et al., 2007.

Reference	Ozone consumption kg O_3/kg TS reduced
Yasui et al. (1996)	0.165
Sakai et al. (1997)	0.133–0.178
Kobayashi et al. (2001)	0.25
Sievers et al. (2004)	0.395
Fabiyi (2007)	0.05–0.07
Rewcastle et al. (2003)	0.13

Note:
kg O_3/kg TS = kilogram ozone per kilogram of total solids

Because cells are lysed by the ozonation process, filamentous bacteria, and therefore foaming, have been observed to decrease. According to Dycztak et al. (2006), sludge volume index (SVI) decreased with O_3 applied to RAS. Fabiyi et al. (2007) observed a decrease in population of *Nocardia* and *Microthrix parvicella* at the Lariana WWTP, Como, Italy.

Many studies cited a TN decrease with sludge ozonation via the production of short-chained VFA being fed to the anoxic zone of a Modified Ludzack-Ettinger (MLE) process after RAS. There has been concern of inhibiting nitrification as a result of RAS ozonation. Bench-scale work conducted by Dytczak et al. (2007) showed that ammonia uptake rates were lower with ozonation, but test and control reactors achieved complete nitrification within the 36-hour HRT with wastewater temperature at 20 ± 1°C. Vergine et al. (2007) reported on a full-scale installation at the Lariana WWTP in Como, Italy, where 62% of the hydraulic load and 75% of the COD load was from textile industries. Ozonation permitted the reduction of WAS by 40% with no decrease in ammonia removal. No change occurred in TN or COD removal of the WWTP. Dewatering of the WAS improved from 19% TS in the cake to 22%.

Process equipment includes the ozone generator, the MLSS side-stream pumping system, and the injection system and contactor. A number of full-scale tests have been conducted by Ondeo-Degremont and Praxair® in Europe, with a couple of recent permanent installations and few more in development. A review of sludge minimization technologies for implementation at the Lebanon WWTP, Oregon, indicated that the Biolysis® "O" process was more costly than other options, such as Cannibal®. However, for industrial wastewater applications, particularly where the footprint is limited, higher doses of ozone required for maximizing WAS reduction may be effective.

2.4.2.2 pH Adjustment

Extreme acidic or alkaline conditions have also been shown to cause cell lysis. Horng et al. (2006) conducted bench-scale tests on polyethylene terephthalate (PET), or polyester, wastewater and compared the impact of acid, alkaline, and thermal hydrolysis on waste sludge, as shown in Table 2-5. Details of the acid and thermal hydrolysis methods were not provided. The alkaline hydrolysis was conducted by addition of sodium hydroxide (NaOH) to raise the pH to 12 for a period of 2 hours. Bench-scale tests were then conducted using parallel trains with a UASB followed by an activated sludge process, each with different amounts of alkaline-hydrolyzed sludge added to the UASB. The greater the volume of hydrolyzed sludge added, the better the performance. Based on this work, a full-scale system was implemented on high-strength wastewater. Addition of the UASB ahead of the existing activated sludge system reduced WAS production 60%. Addition of the hydrolysis step reduced WAS production to 10%.

Table 2-5. Impact of Different Methods of Hydrolysis on Sludge Characteristics. (Horng et al., 2006)

Parameter	Raw Sludge	Acid Hydrolysis	Alkali Hydrolysis	Thermal Hydrolysis
sCOD (mg/L)	160	1,745	2,880	2,970
tCOD (mg/L)	12,470	12,590	12,760	12,540
Hydrolysis rate (%)	–	12.6	21.3	22.4
TSS (mg/L)	12,250	14,680	15,100	12,390
VSS (mg/L)	9,680	9,890	9,320	9,760
NH_4-N (mg/L)	6	31	90	278
TKN (mg/L)	8	52	180	290
TP (mg/L)	0	0.03	0.33	0.55
VFA (mg/L)	61	110	231	216
VSS/TSS	0.79	0.67	0.62	0.79

NOTES:
mg/L = milligram(s) per liter
NH_4-N = ammonium nitrogen
sCOD = soluble chemical oxygen demand
tCOD = total chemical oxygen demand
TKN = total Kjehldahl nitrogen
TP = total phosphorus
TSS = total suspended solids
VFA = volatile fatty acid
VSS = volatile suspended solids

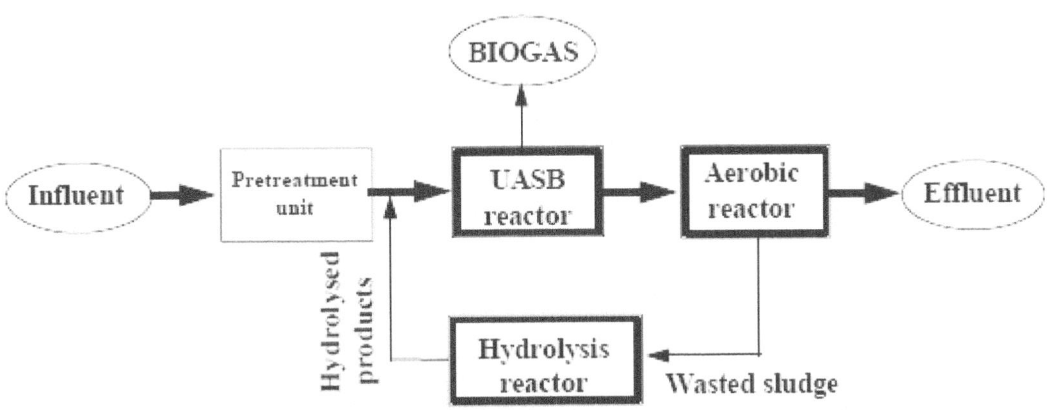

Figure 2-6. Flow Schematic of the Integrated Process with UASB, Aerobic Treatment and Hydrolysis. (Horng et al., 2006)

2.4.2.3 Integrated pH and Temperature Adjustment

Research has also been conducted using a combination of pH and high temperature to promote cell lysis. Based on previous bench-scale work, a large-scale test was conducted on one of the eight bioreactors at the Wabmannsdorf WWTP (Heinzmann and Engel, 2006) using a combination of pH and temperature. The system was designed to treat surface-wasted foam and floating sludge from the bioreactors, with the treated sludge recycled back to the bioreactor. The pH of the sludge was raised to 11 using caustic soda and then maintained at a temperature of 60°C for 2.6 hours before being returned to the bioreactor. The results showed a 17% reduction in WAS production from the test bioreactor, but the goal of reducing floating sludge and filaments in the mixed liquor was not achieved. It was deduced that reduction of filaments would be better served by directing the treated sludge to the anaerobic digesters. Following this work, two other WWTPs had similar chemical and thermal treatment systems installed for treatment of WAS to the digesters using waste heat from the digester gas co-generation systems to heat the sludge to 60-70°C.

2.5 PMC BioTec

PMC BioTec has developed another version of combined chemical and thermal treatment, where chemicals are added on a recirculation loop of a thermophilic digester. Known as the AFC^{sm} process, waste sludges are fed to the thermophilic (45-75°C) bioreactor first. The reactor can either be anaerobic or aerobic via auto thermal aerobic digestion (ATAD). Bioreactor effluent is conveyed to a solids separation process. PMC BioTec has experimented with several solids-separation processes but recommends ultrafiltration. Most of the concentrated solids are returned directly to the thermophilic reactor; the remaining portion goes to chemical treatment. A process flow schematic is shown in Figure 2-7. The process can therefore be described as a catalyzed thermophilic MBR process. In this case, membrane permeate is discharged to the Publicly Owned Treatment Works (POTW) for further treatment and cooling before final discharge. The chemical treatment consists of NaOH addition as well as other chemical treatment. In the ChemTreat reactor, extreme localized ORPs and free-radical production are applied to break down the recalcitrant macromolecules. The chemical treatment step partially solubilizes the excess biosolids before they are returned to the thermophilic reactor for further digestion and destruction. The purpose of the chemical treatment is to enhance biodegradation of solids in the thermophilic reactor. PMC BioTec asserts that the chemical treatment also mitigates foaming in reactors and enhances flux rates of the ultrafilters.

Rozich et al. (2004) performed a four-year study of an aerobic thermophilic biological treatment system that is treating 57 cubic meters per day (m^3/d) of high-strength organic chemical wastes (130,000 mg/L) with over 400 mg/L of toluene and 8% salt (nonvolatile dissolved solids). According to Rozich et al. (2004), COD destruction averaged close to 90%, while effluent toluene levels were all well below effluent requirements (<0.028 mg/L). Less than 3.8 m^3/d of waste sludge is hauled offsite. Influent nonvolatile dissolved solids concentration was lower than the permeate concentration.

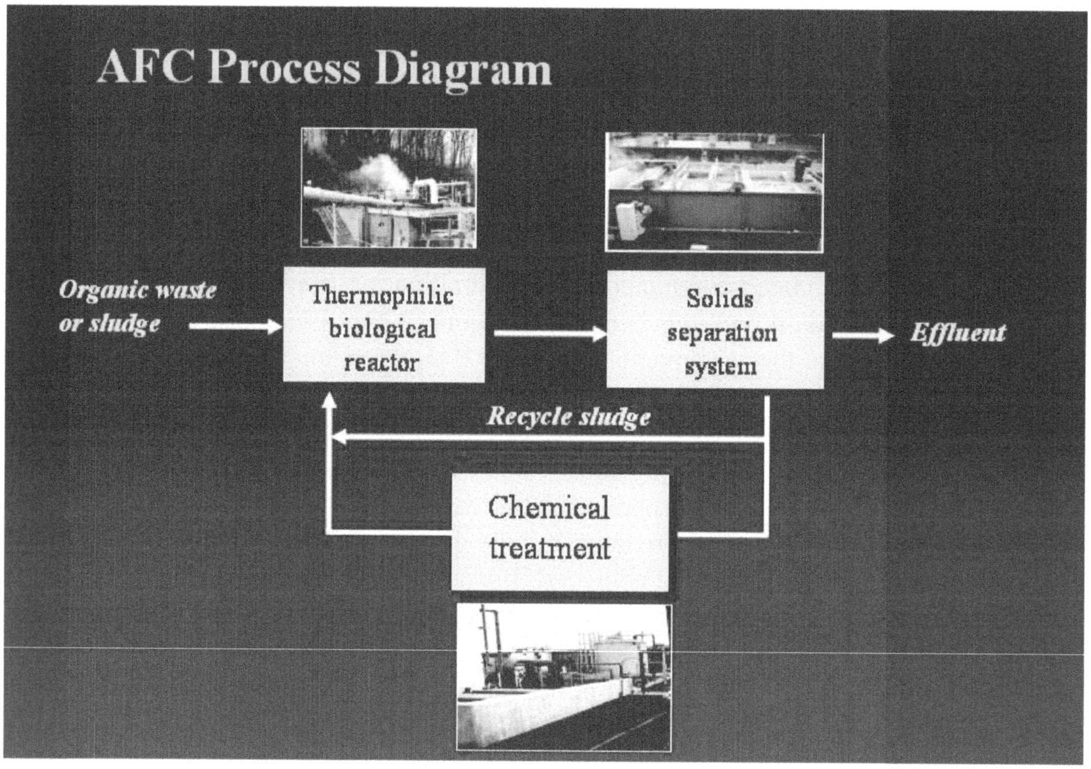

Figure 2-7. Flow Schematic of the PMC BioTec AFC^{sm} Process.
Reprinted with permission from *Proceedings of Water Environment Federation* and the Air & Waste Management *Specialty Conference,* August 22-25 2004, Philadelphia, PA. Copyright © 2004, Water Environment Federation, Alexandria, VA.

In a study reported by Wujcik et al. (2003) at a pharmaceutical plant, a full-scale system was installed after successful bench/pilot testing. Construction of the AFC^{sm} system began in April 2002 and was completed by October 2002. The AFC^{sm} system startup was in October 2002. The system is designed to treat 40,000 pounds COD per day. Claimed results of this study include the following:

- The AFC^{sm} process successfully treated the pharmaceutical wastes.
- The effluent from the treatment system consistently met all discharge requirements.
- COD removal (based on the influent and effluent COD concentrations) averaged 95% since the system startup.
- Pharmaceutical products were not detected in effluent (permeate) samples collected during reactor operation.

♦ No sludge was wasted from the system during the first seven months of operation.

To date, this system has been implemented at full scale at a few industrial facilities. Depending on the strength of the wastewater, the effluent from the system may still require polishing to meet direct discharge criteria, or the effluent may need to be sent to a municipal WWTP for additional treatment. At municipal WWTPs, the process is more likely to be applicable for improving performance of aerobic digesters (ATAD) or anaerobic digesters rather than for direct wastewater treatment.

2.5.1 Chemical Uncoupling

The primary function of suspended and fixed-film biological reactors is to convert soluble organics to suspended solids by creating an environment where biota can use the organic contaminants as a food source. Especially in aerobic environments, an undesired product of this reaction is the creation of excess biota via anabolic metabolism. Carbon dioxide, nitrate, nitrogen gas, and water are other examples of by-products of the cellular metabolism in an aerobic reactor. Anaerobic reactors reduce COD to methane and CO_2. In an anaerobic environment, the cellular growth rate is slower and the metabolic reactions inherently result in lower yields, and therefore less excess biota are produced. Ideally the aerobic reaction would be mostly catabolic with minimal sludge wasting. The notion of uncoupling the metabolic reaction of oxidative phosphorylation is not new. It stems from life sciences – or more specifically, medical- and microbiological-based research. Various chemicals can be used to inhibit metabolism (i.e., oxidative phosphorylation). In a controlled environment with mono-cultures of bacteria, metabolism can be easily inhibited, but this is very different from an activated sludge system, for example, where many different organisms exist and the substrate is highly variable. The concern for wastewater engineers is the potential inhibition of metabolism required to remove the substrate.

According to Okey and Stensel (1993), under energy-sufficient conditions, a discrepancy may occur between the rate of ATP production by catabolism and the rate of ATP use by anabolism – hence, the concept of energy uncoupling. Further, energy uncoupling occurs under chemical inhibitor-present conditions, during transient periods, or at unfavorable temperatures. From an engineering point of view, the concept of energy uncoupling can be extended to the phenomenon that the rate of energy substrate consumption is higher than what would be required for growth and maintenance if the substrate was used efficiently.

Some research has shown that catabolic metabolism can remain functional while inhibiting anabolic metabolism. Cook and Russel (1994) determined that *Streptococcus bovis* dosed with chloraphenol did not reproduce but continued to consume glucose one-third as fast as exponentially growing cells. Cook and Russel (1994) also demonstrated that the nongrowth glucose consumption rate (energy spilling) was 10 times greater than the maintenance rate of growing cells.

Okey and Stensel (1993) determined that addition of the chemicals 2,4 dinitrophenol (DNP) and para-nitrophenol (pNP) were effective at inhibiting anabolism. In a similar study by Strand et al. (1999), it was determined that 2,4,5 trichlorophenol (TCP) and 4-chloro-2-nitrophenol (CNP) were the most effective compounds. In the study by Strand et al. (1999), bench-scale experiments were performed to study the effectiveness. In a continuous feed, complete-mix chemostat, only TCP worked effectively as an uncoupler without inhibiting metabolism. When TCP was fed at high enough concentrations (10 mg/L), it worked effectively. However, tests up to 85 days have shown that activated sludge systems can acclimate to TCP, reducing long-term effectiveness.

To date, no full-scale installations exist. One of the main issues with full-scale implementation of chemical uncoupling is the concern with effluent concentrations of the chemicals and potential environmental impacts, particularly for chlorinated aromatics such as TCP.

2.5.2 Summary

A number of full-scale trials have been conducted of ozonation and pH adjustment processes at municipal and industrial facilities in Europe, with a few permanent installations to date on industrial and municipal applications. The data reported in the literature and by vendors claim high WAS-reduction performance with a number of cases reporting minimal to no sludge wasting. However, initial reviews for the Lebanon Municipal WWTP, Oregon, indicated that ozonation may be less cost effective than other options (Stensel and Strand, 2004). The installations with pH adjustment and pH/temperature processes have been mainly on industrial wastewater. It appears that these systems may be more suited for industrial applications, where solids disposal costs are high, footprint may be restricted, and wastewater strength and temperatures may be high. Reports suggest that a 70-80% reduction of WAS may be achieved without side stream removal of inert material and cell debris. Chemical uncoupling has not yet proven to be applicable at full scale due to toxicity and cost issues.

2.6 Physical Processes

2.6.1 Process Fundamentals and Mechanisms

All physical processes are focused on causing floc disintegration and cell lysis, releasing the organic material in the cells for consumption, either in the secondary treatment process or the sludge digestion process. Most of these processes operate on a continuum from floc disintegration, to cell wall damage, to complete cell disruption, depending on the energy and intensity input. The systems described as follows use different physical methods to cause lysis, from ultrasound and pressure drop processes that cause cavitation, high temperature and pressure, electric pulses, shearing, and homogenization. Some systems can be installed at a number of the different process locations shown in Figure 2-8, while others are more suited for particular locations. All the systems are sized on a volume basis; therefore, the thicker the solids concentration, the more cost effective the system typically is.

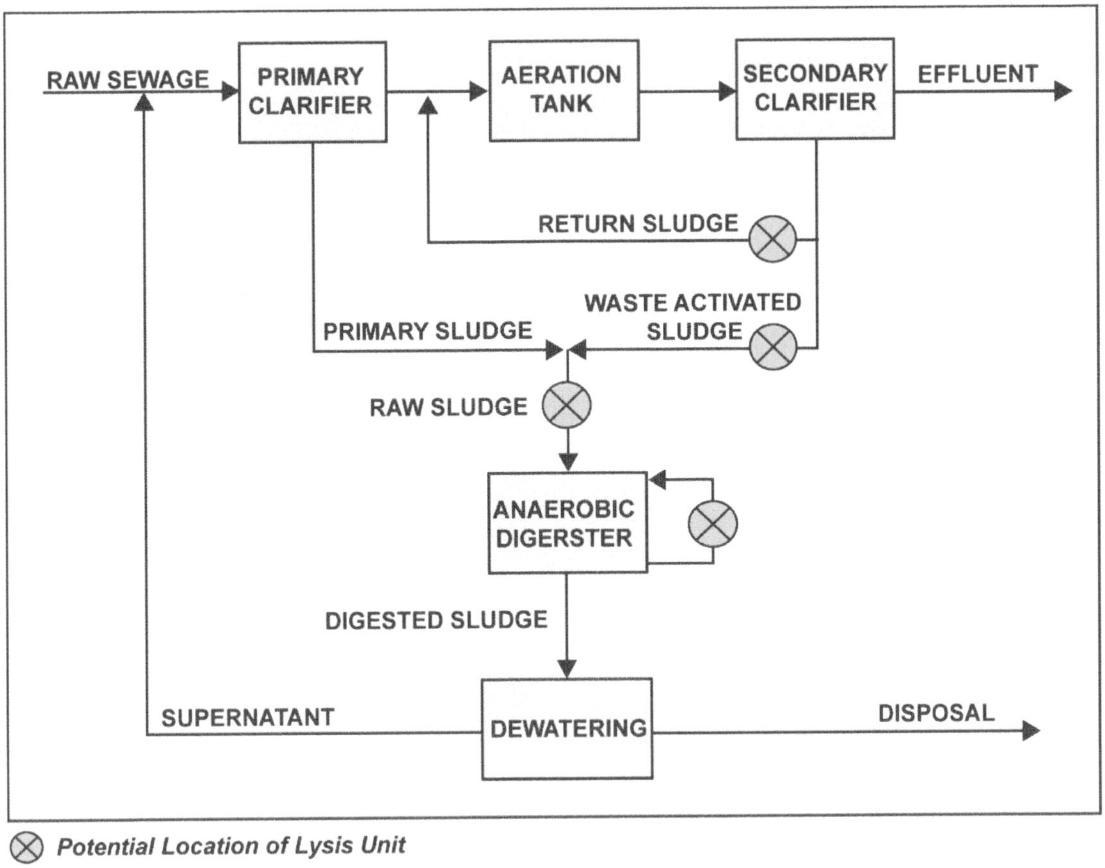

Figure 2-8. Potential Location of Physical Lysis Systems at a WWTP.

2.6.2 Sonication

The application of high-intensity ultrasound has potential for minimizing secondary treatment process WAS production, or improving solids reduction in the digestion process. Ultrasound is sound above the range of human hearing, with frequencies between 20 kilohertz (kHz) and 10 megahertz (MHz) (Neis, 2000; Neis et al. in Mason and Tiehm, 2001; Roxburgh et al., 2005). At the lower end of this range, the compaction and rarefraction waves generated by ultrasound lead to the formation of cavitation bubbles in the fluid, which implode, creating high mechanical shear forces. The implosions create localized hot spots with conditions similar to the sun, reaching temperatures up to 5,000 degrees Kelvin (°K) and pressures up to 500 bar (7,250 pounds per square inch gauge [psig]). Jet streams caused by the implosions can have speeds up to 400 kilometers per hour (250 miles per hour). These forces can be used for disintegrating solids in the fluid. For wastewater applications, it has been shown that ultrasound is most beneficial when applied on biological secondary solids, where rapid hydrolysis can be induced, releasing the nutrients in the cells for consumption in the activated sludge or anaerobic digestion process.

Work by the Technical University of Hamburg Harburg (Neis et al. in Mason and Tiehm, 2001) and the Fraunhofer Institute in Germany, Atkins Water in the U.K., and other university research, such as the recent work by Muller et al. (2005), Andreottola et al. (2006), and Khanal et al. (2006), has shown ultrasound operates on a continuum from disintegration of flocs through cell wall damage to complete cell disruption, depending on treatment conditions such as the input power and intensity of energy applied. Floc disintegration increases surface area and may release extracellular polymeric substances (ESP) and activate enzymatic activity, while cell disruption increases sCOD and VFA concentrations. Muller et al. (2005) analyzed the cations in solution during bench-scale ultrasound digestion tests. The data showed minimal differences between solution cations (sodium, potassium, magnesium, and calcium) in the test and control digesters. It has been proposed (Higgins and Novak, 1997) that calcium and magnesium serve as ionic bridges between lectin-like proteins secreted by flocculent bacteria. The data suggest that under the sonication conditions in the tests conducted by Muller et al. (2005), lectin-like matter was not solubilized.

Ultrasound is used in a wide range of industrial applications, in many different frequencies and forms. Therefore, an extensive list of ultrasound manufacturers exists in North America and Europe. However, the number of manufacturers with equipment designed for wastewater treatment applications and with experience in this field is limited. The two main suppliers are UltraWaves, represented in North America by Dorr Oliver Eimco®'s Sonolyzer™ equipment (Vera et al., 2005), and Sonico with the sonix™ system. UltraWaves is a German company that was established as a spinoff from ultrasound research conducted at the Technical University of Hamburg-Harburg, which purchases the ultrasound equipment from Sonotronic, a local manufacturer. Sonico is a U.K. company that was established as a joint venture company between Purac, an equipment supply company, and Atkins, an engineering consulting company that conducted applied research in ultrasound wastewater applications. Sonico purchases the ultrasound equipment from Branson, a company based in the U.S. These two companies have the largest number of installations in wastewater applications, with over 30 installations in Europe, Asia, and Australia (Kruger and Hogan, 2005; Sonico company installation list; UltraWaves company installation list). The largest installation is the Sonico installation for WAS pretreatment for enhanced digestion at the 400-MLD Mangere WWTP in New Zealand. Two smaller companies also have a few smaller wastewater installations, Ultrasonus from Sweden and VTA Technologie GMBH from Austria. IWE Tec was a German company with several full-scale installations in Germany that had acquired representation in North America in 2003; however, this company has since ceased operation.

Ultrasound may be generated by two different methods – magnetostrictive and piezoelectric. The former uses electric energy, passing through a magnetic coil attached to the vibrating piece to produce the mechanical energy, or vibration. The latter uses electrical energy, converted to high-frequency electric energy, which is applied to piezoelectric crystals that vibrate at the same frequency. The crystals are attached to the vibrating piece (known as the sonotrode, probe, or horn), causing the vibration to be transferred to the liquid. Magnetostrictive systems typically have a longer life but lower energy efficiency, as the electrical energy applied is converted to magnetic energy prior to being converted to mechanical energy. For wastewater applications, it appears that the economics favor the use of piezoelectric systems due to the high-energy intensity required to lyse the cellular material in the sludge.

Differences exist in the ultrasound systems available for wastewater treatment. Each manufacturer uses a unique shape for the vibrating ultrasound piece (e.g., UltraWaves uses short rod-shaped sonotrodes that project into the flow path [Figure 2-9], while Sonico uses ring-shaped horns that sit within a pipe spool [Figure 2-10]). UltraWaves treats a smaller portion of the flow for a longer retention time, while Sonico typically treats a greater portion of the sludge flow for a considerably shorter retention time. Each probe in the UltraWaves system is rated for 2 kilowatt (kW) power input but typically operates at 1 kW. Sonico has 3 and 6 kW horns, with the latter as their standard unit, operating at 50-60% of rated power.

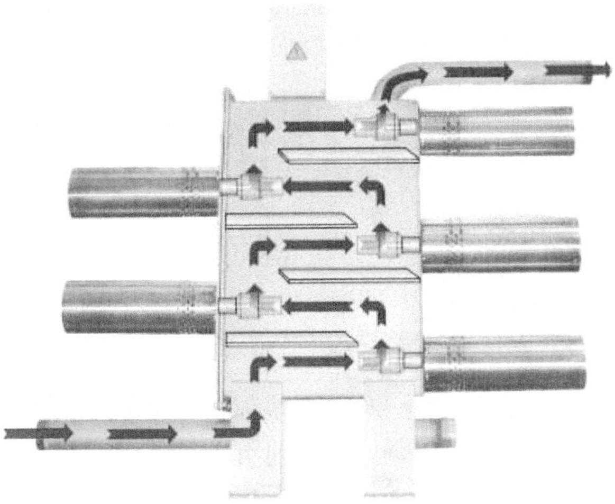

Figure 2-9. UltraWaves Sonolyzer™ Unit.
Reprinted with permission.

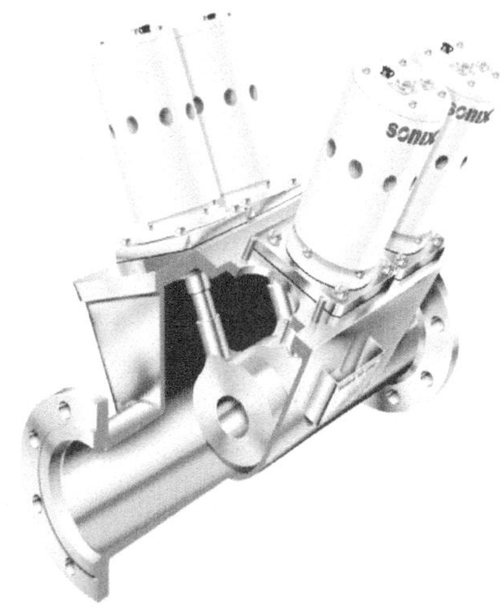

Figure 2-10. Sonico Ultrasound Unit.
Reprinted with permission.

The majority of ultrasound applications to date have been for WAS pretreatment prior to anaerobic digestion. However, both UltraWaves and Sonico have European installations on RAS streams within the WAS process.

Ultrasound installations in Europe show that treating a side stream of the RAS flow can reduce WAS production from the secondary treatment process by 25-50%, depending on characteristics of the sludge, operation of the secondary treatment process, and the amount of power input to the system. Sludge minimization tests were conducted by Anglian Water in the U.K. to compare bioaugmentation (addition of a fermented bacteria and nutrient solution) with a Sonico ultrasound system in parallel activated sludge lanes (Griffiths, 2001). The control lane was operated at an SRT of 12 to 15 days and a mixed liquor concentration of 3,500 mg/L. The test showed that a titanium ultrasound horn system treating approximately one-sixth of the RAS flow provided up to 30% reduction in WAS production, compared with 5 to 9% reduction with the bioaugmentation process. The ultrasound system was also found to have an additional advantage in reducing the impact of filamentous foaming and maintaining good settleability (95 milliliters per gallon [mL/g]) during a period when filamentous organisms reduced the settleability of the control lanes (140 mL/g). No increase in aeration demand was noted during the test.

The Leinetal WWTP in Heiligenstadt, Germany, tested an UltraWaves ultrasound system prior to full-scale installation. The activated sludge process was operated as an extended aeration process, with an SRT of 18 days, and experienced problems associated with bulking sludge. The ultrasound system was installed to treat 30% of the thickened secondary sludge flow, which was then returned to the aeration basins. The test showed that the ultrasound system reduced WAS production by 30%, improved settleability of the secondary sludge, and increased the dewatered cake dryness by two percentage points. Foaming and bulking problems were also eliminated. The full-scale installation has been operational since June 2003, and the plant has been able to avoid construction of a new aeration tank (Eimco/UltraWaves case study sheet).

Treating a thickened sludge stream, rather than unthickened RAS, may improve the cost effectiveness of ultrasound installations for activated sludge applications, as the ultrasound systems are designed to maintain a specified hydraulic retention time within the sonication chamber to achieve cell lysis. However, around 6 to 7% solids concentration the energy required to overcome the sonication threshold or cavitation attenuation caused by higher viscosities and solids concentrations will rise.

Ultrasound applications for WAS pretreatment prior to anaerobic digestion have shown that with appropriate levels of sonication, WAS will achieve similar digestion rates to PS, typically increasing the VSr of the WAS from 40 to 60%. The overall improvement in digester VSr will depend on the ratio of primary to WAS in the digester feed, as well as other site-specific conditions, such as baseline performance of the digesters and sludge characteristics. As ultrasound increases the hydrolysis rate, the impact will be greatest for digesters at short HRTs where cell lysis is rate limiting. Bench-scale work by Muller et al. (2005) also showed that the use of ultrasound reduced the organo-sulfur odor potential with a 10 to 24% reduction in headspace sulfur compounds compared with the control.

Effectiveness of the ultrasound system can be affected by a number of factors. Excess polymer dosing for WAS thickening will reduce the effectiveness of ultrasound as particles may re-agglomerate if there is active polymer still present. Pilot tests conducted in the U.S. have been at plants in southern California, which add ferric chloride for odor and digester gas hydrogen sulphide control. Metallurgy of the ultrasound equipment immersed in the sludge is such that chloride concentrations above 200 mg/L increase the wear on the probes, impacting life and effectiveness of the system. Analyses on WAS from the Riverside Water Pollution Control Plant, California, had chloride concentrations as high as 1,200 mg/L due to addition of ferric chloride and chlorine for odor control in the headworks and primary treatment processes (CH2M HILL, 2005). Later work by Paradigm Environmental Technologies when testing the MicroSludge® process at the Joint Water Pollution Control Plant (JWPCP) in Los Angeles County has shown that ferric chloride may inhibit the WAS digestion and reduce the effectiveness of cell lysis, which may also have been responsible for lower-than-expected performance when ultrasound was tested at the same plant.

2.6.3 Thermal Hydrolysis

Thermal hydrolysis uses temperature and pressure to cause hydrolysis and break open cellular material. Research into thermal hydrolysis for improved anaerobic digestion was started in the 1970s and was considered in process evaluation conducted by CH2M HILL for the City of Los Angeles in 1977. However, at that time, the technology faced a number of operational and maintenance problems, including high-pressure pumping of the feed sludge, fouling of heat exchangers, and wear on equipment. In the 1990s, thermal hydrolysis was evaluated in Norway, and has since developed in to a viable technology, using newer technology, such as direct steam injection rather than heat exchangers. This process has received considerable attention in Europe, primarily through the efforts of Cambi™ AS, a company that grew out of the research in Norway. In the U.S., pilot-scale work has been done by the San Francisco Public Utilities Commission (SFPUC), using a 2-liter thermal hydrolysis unit (Panter et al., 2006). The systems that have been commercialized to date are based on a batch operation mode. Continuous thermal hydrolysis systems are being developed but are in the early stages. Due to the high temperature and pressure conditions that occur in thermal hydrolysis, the cost of the equipment is high and means that demonstration-scale equipment requires a significant investment and may not be cost effective if it is done at a small scale.

The technology may be retrofitted ahead of existing digesters or may be incorporated into new facilities. Key advantages of thermal hydrolysis include the ability to operate digesters at high-feed solids concentrations, of around 10%, due to improvements in viscosity through the process, improved biosolids dewaterability, and the ability to achieve Class A pathogen reduction through operation of a batch treatment process. Hydrolysis of the feed solids provides for improved gas production and solids destruction. However, the cost of high-pressure stainless steel tanks, heating, and the need for operators trained in high-pressure systems has impacted the cost and the acceptability of this process in North America. It has been realized that the cost of the system and some of the complexity may be reduced by treating only the WAS, rather than the PS and WAS combined. This provides some important advantages over the previous approach, including the following:

- Smaller system, as only the WAS is treated reduces costs.
- Heat recovery heat exchangers may be avoided, as the heated WAS can be used to heat the PS.
- Treating only the WAS focuses the costs on the portion of sludge with the highest return for improved digestion.

2.6.3.1 The Cambi™ Thermal Hydrolysis Process

The continuous steam hydrolysis is in the early stages of development; this process review will focus on the Cambi™ batch system. Thirteen Cambi™ thermal hydrolysis process (THP) installations are in place in Europe. In the U.S., Cambi™ AS has retained RDP Technologies Inc. (RDP) as its sales representative.

In the Cambi™ THP process, shown in Figure 2-11, a three-tank system is typical. Pre-dewatered solids, up to 15% solids concentration, are added to the feed tank. The feed tank is heated by the steam from the reactor and the flash tank. The solids are held in the feed tank until the reactor is ready for a batch. Solids are pumped to the reactor where they are held for a minimum of 20 minutes at 170°C and 8 bar pressure. Live steam is added to achieve this temperature and pressure. Steam is vented to the feed tank. At the end of the batch, solids are allowed to flow to the flash tank where the rapid expansion causes the cells to rupture and the resulting solids have considerably lower viscosity. The temperature is also reduced by flashing, so the temperature in the digester is about 37°C. By the addition of the water in the steam and the cell rupture, the resulting solids concentration to digestion is 10-12%. Due to the cell rupture, the solids are now easily pumped to the digestion process.

Normally, the solids in the digester will require no further heating. Although the Cambi™ THP process requires considerable heat for steam, some heat is recovered in the process and is used for digestion. To improve the heat balance of the process, an alternative approach has been proposed for applications that do not require Class A pathogen requirements to be met. Applications of Cambi™ THP in Europe treat the entire solids stream, including primary and secondary sludges. However, the main hydrolysis benefit is on the secondary sludges. The cost effectiveness of the system could be improved by applying thermal hydrolysis only to the secondary sludge. This would reduce the tankage and the steam required. The secondary sludge could then be mixed with unheated PS to bring the combined feed temperature to that required for mesophilic digestion.

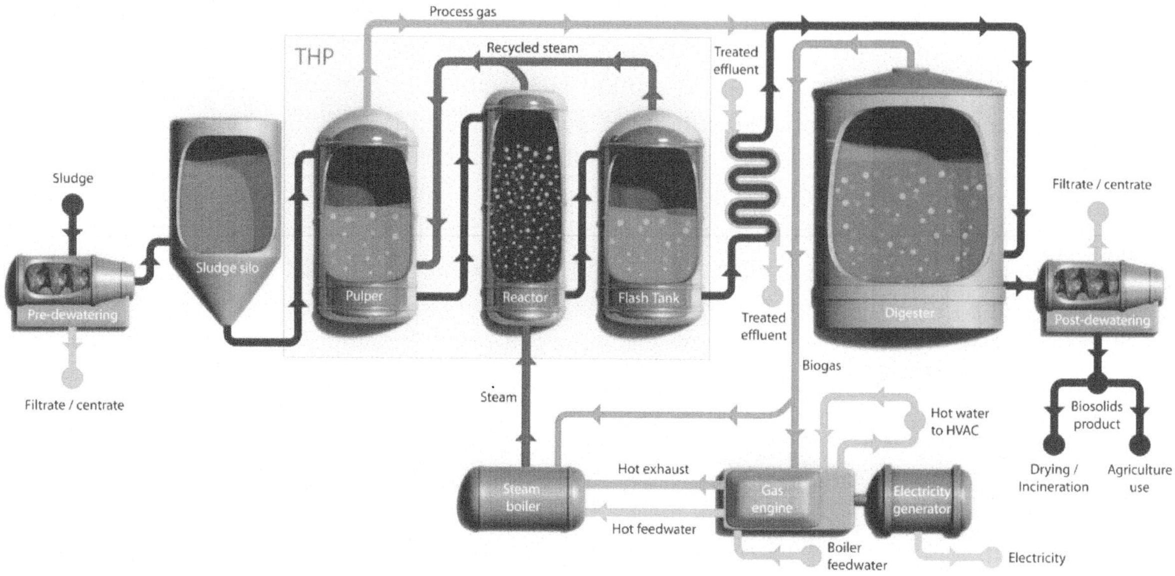

Figure 2-11. Cambi™ THP Schematic.
Reprinted with permission from Cambi.

Tests conducted in Germany on WAS-only hydrolysis indicate that digesters could be fed at 8% TS combined WAS and PS, depending on viscosity and PS thickening efficiency.

Proteins and exopolymeric substances (EPS) have also been linked to reduced dewaterability. Thermal treatment above 100°C has been shown to denature proteins and EPS. The higher temperature and pressure of the Cambi™ THP denatures and dissolves proteins, improving dewaterability (Kopp et al., 2007). Bench-scale tests on hydrolyzed WAS combined with untreated PS for anaerobic digestion showed a 25% increase in gas production and 8 percentage points improvement in dewatered solids content compared with the control, although the polymer demand was higher with the Cambi™ THP.

2.6.3.2 The *Bio*Thelys® Process

A similar thermal hydrolysis process has been commercialized by Veolia and installed in 2006 at two cities in France – Saumur (60,000 pe) and Chateau-Gontier (38,000 pe). A process schematic is provided in Figure 2-12. The Saumur WWTP is an extended aeration process without primary treatment. Results from the *Bio*Thelys® process were documented by Chauzy et al. (2007). As the digester was started up in August 2006, the data reported are very preliminary, and VSr information was not provided. However, early dewatering data indicate that centrifuge dewatering would achieve close to 30% solids in the cake.

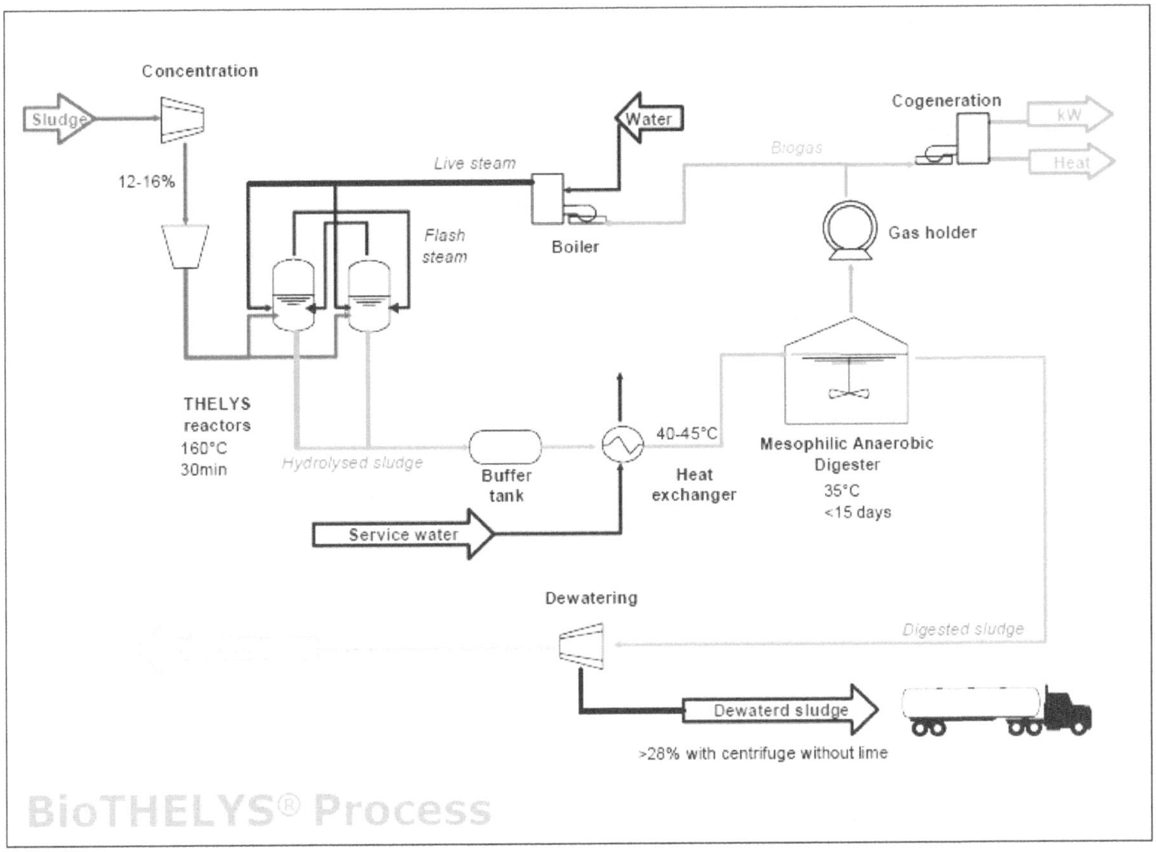

Figure 2-12. The Veolia BioThelys® Process Schematic.
Reprinted from *Water Practice and Technology*, 3:1 (2008), with permission from IWA Publishing.

2.6.3.3 The SUBBOR Process

The Super Blue Box Recycling Corporation has developed the SUBBOR steam explosion process for improved digestion of municipal solid waste. The system is designed to pretreat wastes at 30% total solids concentration using steam to heat the material (180-260ºC) and pressurize the material (1 to 40 bar), followed by rapid pressure release through a proprietary orifice. Dereix et al. (2005) reported on tests conducted with a pilot-scale cooker to determine the effectiveness of the SUBBOR process on municipal WAS. Due to limitations in obtaining WAS samples thickened to the required concentration, WAS was mixed with dewatered biosolids from anaerobic digesters at the City of Ottawa's Robert O. Pickard Environmental Centre (ROPEC) to produce a mixture at 15% TS concentration for steam explosion. Bench-scale digestion tests showed that at a treatment pressure of 20 bar gas production in the thickened WAS (TWAS)/biosolids mixture was 50% higher than the control. Dewaterability was measured using the capillary suction test and showed a 32% improvement. Under the test conditions, the excess biogas generated was not adequate to provide the heat required for steam generation. However, if the process could be optimized for higher TWAS solids and tested without addition of digested biosolids, the economics of the process would be improved.

2.6.3.4 Microwave Pretreatment

Microwave pretreatment to improve anaerobic digestion has been tested at the bench scale. Work conducted at the University of Ottawa, Canada, has shown that microwave irradiation can damage flocs and cell membranes, releasing exocellular and intracellular material, and can also solubilize particulate COD. Bench-scale digestion tests were conducted by Eskicioglu et al. (2006) to determine whether microwave heating had any athermal advantages over conventional thermal heating. Tests were conducted at 50 and 96°C using microwave and conventional heating methods and compared with untreated controls. The data reported indicate that conventional heating of WAS resulted in better digestion and dewatering performance than microwave heating, while both heating methods performed better than the control.

2.6.4 Homogenization – MicroSludge®

The MicroSludge® process is a sludge pretreatment process patented by Stephenson and Dhaliwal in 2000 and developed by Paradigm Environmental Technologies, Vancouver, British Columbia, Canada. The process changes both the rate and the extent that WAS from a secondary WWTP is degraded in a conventional mesophilic anaerobic digester. It uses alkaline pretreatment to weaken cell membranes and lower viscosity and an industrial-scale, high-pressure homogenizer to provide an enormous and sudden pressure drop to burst the bacterial cells in the sludge. The resulting liquefied WAS is anaerobically digested together with the thickened PS to form stabilized biosolids and biogas.

The MicroSludge® process uses alkaline pretreatment to weaken cell membranes and reduce viscosity, mechanical shear to reduce particle size, a self-cleaning screen to remove noncellular oversize debris, and an industrial-scale homogenizer to provide an enormous and sudden pressure change to burst the cells. The central unit of the MicroSludge® process is an industrial-scale homogenizer that provides a large and abrupt pressure drop. At 82,700 kilopascals (kPa), WAS in the cell disruption homogenizing valve is accelerated up to 305 meters per second, nominally the speed of sound, in about two microseconds (Pandolfe, 1993). This high-velocity flow causes enormous fluid shear and cavitation and then impinges on an impact ring, disrupting the cell membranes and producing a liquefied WAS homogenate (Middleberg, 1995).

The chemical step uses sodium hydroxide to raise the pH between 9 and 10. Chopper pumps are used to break up large particles, and a fine 800-µm screen is used to remove particles that could damage the homogenizer. A gas-liquid separator and pulsation dampener is also required prior to the homogenizer to prevent gas-locks occurring. The homogenizer itself is essentially a valve with a narrow passageway of less than 1 millimeter (mm), through which the sludge is pumped at 82,700 kPa. The pressure drop across the homogenizer liquefies the sludge and also provides a net increase in sludge temperature of 20-25°C. The principal operation and maintenance (O&M) costs of the process are chemicals and power. The principal benefits of the process are the increased biogas production associated with the increased VSr, a reduction in the digester HRT, and the reduced quantity of digested biosolids that require dewatering and disposal. To date this process has been tested for WAS pretreatment for enhanced anaerobic digestion. It is conceivable that the process could equally be applied on a side stream of RAS to provide reduced sludge production from the activated sludge process.

A series of laboratory and pilot scale tests of the MicroSludge® process was conducted between 2000 and 2003. The process proved to be highly effective in conditioning WAS prior to anaerobic digestion. For example, the volatile solid levels in MicroSludge® pretreated WAS was reduced by 68% after 15 days of mesophilic anaerobic digestion. Normally, conventional mesophilic anaerobic digestion of unconditioned WAS results in approximately 30% VSr after 15 days. MicroSludge® pretreatment also led to a net 96% reduction of the influent VFA concentrations of the sludge during digestion, removing the readily biodegradable portion and correspondingly reducing its odor potential.

The first full-scale MicroSludge® plant was commissioned in January 2004 at the Chilliwack WWTP, approximately 115 km east of Vancouver, British Columbia, Canada. The Chilliwack WWTP serves a population of 70,000, and a single 4,000-liter-per-hour (L/h) homogenizer operated for approximately 10 hours per day to pretreat all of the TWAS generated at this facility. The treated homogenate was stored in a holding tank and fed to the digesters at a continuous rate. The digester retention time was 13 days. This facility operated as a demonstration for approximately three months. The results at Chilliwack indicated greater than 85% VSr of the WAS treated (Stephenson et al., 2005). However, it was not possible to confirm the digester mass balance as the gas production was not measured, nor were the dewatered biosolids.

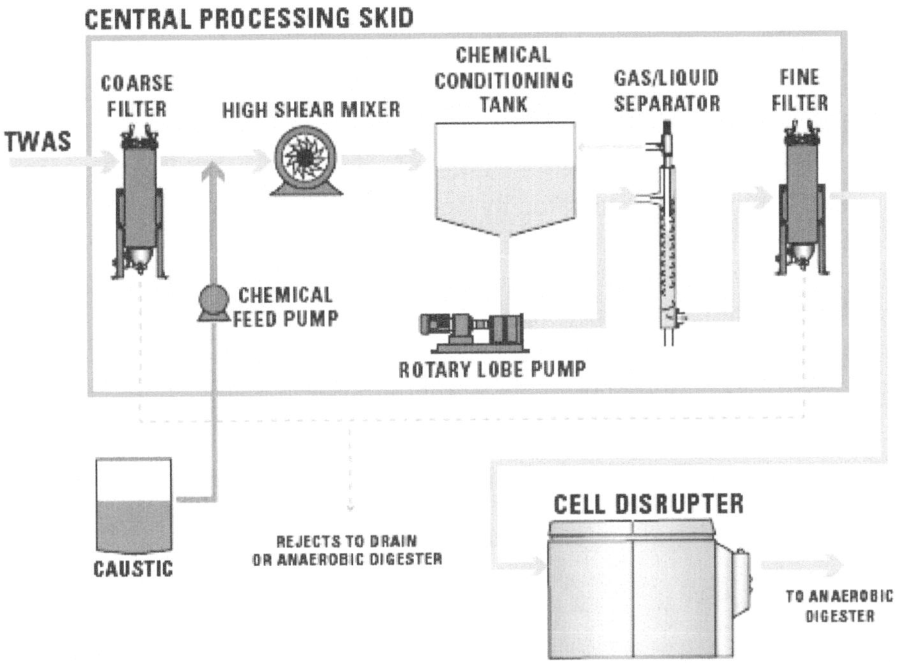

Figure 2-13. MicroSludge® Process Flow Schematic. (Courtesy Paradigm Environmental)

In 2005 and 2006, MicroSludge® operated a demonstration at the Los Angeles County JWPCP, California, with two 4,000-L/h homogenizers processing TWAS at 5-6% followed by blending with PS in a PS:TWAS ratio of 75:25 on a volumetric basis and digested for 19 days. This facility did not have a homogenate holding tank. Modest increases in VSr and gas production were noted compared with the control digester. This was attributed to inhibition by high levels of ferric chloride in the PS (Rabinowitz and Stephenson, 2006), where iron salt was added for control of hydrogen sulfide by precipitation as iron sulfide. Speece et al. (2005) have reported that methanogenic enzymes comprise approximately 40% sulfur; therefore, methanogens have a high obligate requirement for sulfide. However, sludge and dewatered cake odor tests indicate that odor potential, as measured by total volatile organic sulfur compounds (TVOSC), was significantly reduced (up to 50%) by the MicroSludge® process compared with the control digester (Novak et al., 2007)

Paradigm completed another full-scale demonstration at a WWTP in Des Moines, Iowa, in the middle of 2009. An evaluation of the data collected during this effort is presented in a separate report.

2.6.5 Pressure Release

The Crown® Disintegration System was recently acquired by Siemens. The Crown® disintegration system is a multiple-stage disintegration system without chemical preconditioning. It may be used on a RAS side stream to reduce sludge production from the secondary treatment process or on the WAS feed to improve digestion.

The system components consist of the following:

- Homogenizer or macerator
- Crown® disintegrator nozzle
- Control system
- Pressurization pump
- Recirculation tank
- Discharge pump

The Crown® disintegration system uses the principle of a carefully controlled cavitation process. This type of cavitation is generated by operating the system at 12 bar and pumping through the disintegrator, which is a pressure reduction nozzle. Cavitation occurs in the second part of the nozzle due to the sudden pressure drop. Between 30 and 100% of the WAS flow may be treated through the system. Typically, the WAS is cycled round through the disintegrator three times. For anaerobic digester applications, the vendor guarantees at least 20% increase in gas production and 15% reduction in dry solids volume.

Figure 2-14. Crown® Disintegration System at Rosedale WWTP, North Shore, New Zealand.

The Crown® installation list includes 17 permanent facilities. Most of the existing installations are in Germany. The Crown® system was also installed at North Shore City's

Rosedale WWTP in New Zealand. North Shore City decided to install a cell lysis system to improve digester and control foaming. As most cell lysis systems are sized on flow rate, to reduce the size and cost of the system a WAS thickening system was included as part of the bid package. The bid was opened to Sonico, IWE Tec, MicroSludge®, and Crown® Biogest. Flo-Dry Ltd., the local representative of the Crown® system, won the project and installed a gravity belt thickener and the disintegration system. Thickening increased the digester retention time from 17 to 24 days. The system met the performance requirements to achieve 23% increase in gas production (Susarla et al., 2006). Some initial problems presented themselves, primarily related to pumping the sludge when WAS TS concentration was higher than 6.5%. The system has been operating almost a year, and the City staff intends to shut down the system soon to examine wear and tear on the equipment. The disintegrator is warranted for 18 months. The City is also planning to leave the Crown® system offline for a few months to assist in evaluating how much of the improved digestion can be attributed to the Crown® system and how much is due to the increased digester retention time.

2.6.6 Pulsed Electric Field

Cell disruption using electrical pulses has been in use in the food and other industries for decades, but earlier vendors have had little success in the wastewater industry. Pulsed electric field, or electron beam systems, use pulsed electric currents within the sludge stream to cause shock waves and induce cell lysis. The technology was initially developed in Germany in 1997 and licensed to Scientific Utilization Inc. (SUI), a U.S. company founded in 1992. The process operates by immersion of two electrodes into the sludge stream. Initial tests in Germany for WAS pretreatment prior to anaerobic digestion indicated that a gas production increase of 12 to 20% could be achievable (STOWA, 2003a). The process was tested using the SUI PulsePower™ submerged plasma arc technology at a number of installations in the U.S., including OCSD, California. Results indicated that the input energy was not sufficient to provide significant improvements in WAS reduction through the digestion process. In part due to lack of funding to further develop the technology, this application was not pursued further by SUI. Further information on the SUI technology can be found in a WERF report by Dentel et al. (2000).

The technology appears to have given rise to recent interest from new companies that are seeking to develop similar processes, including OpenCEL, based out of Chicago, U.S., and EB-Tech, based in South Korea.

2.6.6.1 OpenCEL

OpenCEL is an environmental biotechnology company with a proprietary process, Focused Pulse (FP), for the treatment of biological solids produced in municipal and industrial wastewater processing. The OpenCEL technology treats biological solids by first subjecting them to high-voltage electrical micropulses, typically between 20 and 60 kilovolts (kV) per centimeter, and then allowing the damaged cells to biodegrade. The microbursts cause electroporation or lysis of the cellular membranes, where the cells are exposed to high-voltage pulses that temporarily destabilize the lipid layer and proteins of cell membranes. The plasma membranes of the cells become permeable to small molecules after being exposed to an electric field, and permeation then causes swelling and eventual rupture of the cell membrane. This treatment is conducted at ambient temperature, and the total treatment of the cellular matter is on the order of 20 to 100 microseconds.

The OpenCEL technology incorporates commercially available pulse generation equipment, consisting of step-up voltage transformers, capacitors for electrical energy storage, switching equipment that can generate micropulses of energy discharge, and a treatment chamber where an electrode assembly focuses the microburst on the biosolids. The equipment occupies minimal footprint, has no moving parts, and has little or no wear components. Figure 2-15 provides a schematic of the system.

Extensive testing has been carried out at the Center for Environmental Biotechnology at Arizona State University. An initial trial was carried out on aerobic wastewater at Lancaster WWTP in Lancaster, Ohio (Banaszak and Klein, 2007). Approximately 33% of the return flow was treated and fed into the aeration basins. The net volume of solids requiring disposal was reduced by 24% for treatment of 33% of the total RAS flow.

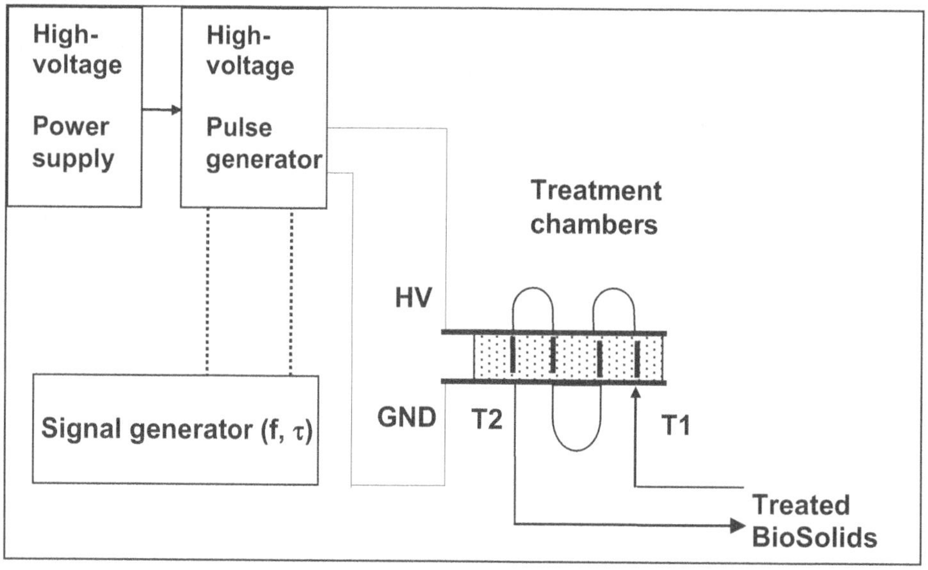

Figure 2-15. OpenCEL Focused Pulse System Schematic. (Banaszak and Klein, 2007).

A commercial, full-scale demonstration of the OpenCEL technology is being conducted at the Mesa, Arizona, Northwest Water Reclamation Plant (NWWRP) to focus on determining the net energy gain and associated waste biosolids reduction attained by applying OpenCEL's technology. The plant can process up to 5.6 liters per second (L/s) of blended thickened sludge and will typically operate at about 2.8 L/s.

2.6.7 Shearing

2.6.7.1 ABS Kady

Not to be confused with the IDI Biolysis® systems previously described, the ABS-Kady® Biolysis® Process was developed as an offshoot application by Kady® International, which manufactures mills for industrial applications. The physical shearing action provided by the Kady® high-speed rotor-stator mill can be used to promote cell lysis of biological sludge, although it appeared that the mill itself did not lyse cells but reduced particle size and broke up filaments. A number of tests were conducted at different locations in the treatment process, including activated sludge recycle streams, WAS pretreatment for enhanced digestion, and digester recirculation loops, with varying success.

The system was tested for almost a year at the 254 MLD Springfield Regional Wastewater Treatment Facility (WWTF), Massachusetts, as part of a process to improve economics of the wastewater treatment process (Borgatti, 2004). The activated sludge system was changed to an extended aeration system, with a SRT of 18-22 days and a MLSS concentration increase from 2,000 to 5,000 mg/L. The Kady® process was installed on two of the four trains, with thickened sludge pumped through the mill and returned to the secondary treatment basins. The trial results indicated that the extended aeration process provided a 30% reduction in WAS production compared with the lower SRT activated sludge process, while addition of the Kady® process to two of the extended aeration trains increased the WAS reduction to 42%, providing a 40% improvement over the extended aeration system. The extended aeration process increased the aeration demand for the activated sludge process, but the demand in the Kady® process trains was 8-9% lower than the other two trains. There were associated cost reductions in downstream solids processing and handling. However, the Kady® process did not appear to improve the SVI and appeared to be susceptible to settling problems due to spring wet weather flows, and these issues were not resolved at the time.

The Kady® Biolysis® system was also tested on the digester heat recirculation loop at the 20.8 MLD Caldwell WWTP, Idaho (Basu et al., 2004). The digester was fed primary and secondary sludges. A grinder and basket strainer were installed upstream of the mill. Sludge loading to a test digester was increased in increments during the trial. A similar digester was used as the "control" but was maintained at the baseline loading rate. However, the average VSr in the control digester appeared to be uncommonly low, at 32% given a nominal HRT close to 24 days. This VSr was lower than the US EPA *Code of Federal Regulations* (CFR) 503 regulation VSr of 38% for Class B biosolids. Volatile solids reduction in the test digester averaged 60%. Gas production from each digester was not monitored; therefore, the difference in VSr cannot be verified by this means.

Bench-scale tests have also been conducted for Georgia-Pacific on pulp and paper treatment comparing the Kady® Biolysis® process with and without chemical preconditioning with a control system. The OUR was measured to indicate soluble BOD release through the different configurations (ABS-Kady® International, 2006).

Due to restructuring within the company and a decision to focus on core business areas, the Kady wastewater process was withdrawn from the market. Recent communications appear to indicate that new supply agreements may bring the system back on the market.

2.6.7.2 Lysate Centrifuge

The lysate centrifuge was developed in Germany and was known as the Baker Process, as mergers between German company and Baker Hughes in the U.S. led to the process being assigned to the Baker Hughes environmental division. The technology incorporates a lysate ring or rotating cutter, in the thickened sludge discharge of the centrifuge to shear cellular material, using the energy inherent in the discharge stream. A number of pilot tests were conducted in Germany and the Czech Republic in the late 1990s and early 2000s. These tests focused on sludge pretreatment for enhanced anaerobic digestion. Results showed digestion improvements, with up to 25% increase in digester gas production (STOWA, 2003b). However, Baker Hughes has since sold or dissolved its environmental division. The Bird Humboldt centrifuge business was bought by Andritz, but the Baker Process was not included in the sale. Dr. Dohanyos and his group at the Institute of Chemical Technology in Prague have conducted research and implemented full-scale installations using the lysate centrifuge technology. A recent paper by

Zabranska et al. (2006) summarizes the work conducted to date and presents data from full-scale installations at two municipal WWTPs in the Czech Republic and two in Germany. Three of the facilities have the lysate centrifuge technology installed with mesophilic digestion. Increase in gas production and VSr has been in the range of 15-26%. The Central WWTP in Prague had the lysate centrifuge technology installed, followed by conversion to thermophilic digestion. The combined effect has been a 38% increase in the specific gas yield (m^3/kg volatile solids in).

2.6.8 Summary

A number of physical cell lysis systems are in operation in Europe and a few in Asia, Australia, and New Zealand, ranging from over 30 ultrasound installations on digesters or secondary treatment trains, 17 Crown® Biogest® installations on digesters or secondary treatment trains, four lysate centrifuge installations, and around 10 thermal hydrolysis installations on anaerobic digesters, of which two are at plants that do not produce PS. Data indicate that physical lysis when installed on the RAS line in the secondary treatment process may achieve up to 30% reduction in WAS production. In digester pretreatment applications, the WAS digestability can be improved to a level similar to PS; therefore, the overall digester improvement will depend on the ratio of primary to WAS. The Cambi™ THP installation in Naestved, Denmark, which did not produce PS, provided 60% improvement in VSr and improved dewaterability (Fjordside, 2002). Bench-scale data have shown that conventional thermal pretreatment performs better than microwave pretreatment.

In North America, a number of large-scale tests have been conducted with ultrasound, MicroSludge® homogenization, pulse power, and shearing with the Kady® mill. However, to date no full-scale permanent installations of any physical technology are located in North America, apart from the OpenCEL facility in Mesa, Arizona.

2.7 Conclusions

The state of development of sludge minimization technologies continues to advance, both for applications within the liquid treatment process and for digester pretreatment. Most technologies have full-scale applications on industrial and municipal wastewater in Europe and Australasia, with a much slower implementation rate in North America. The exceptions to this trend are the combination biological processes such as Cannibal®, which have been developed in North America and found wider acceptance. However, many technologies have a limited number of installations (e.g., chemical treatment technologies), and some technologies have shown mixed performance, particularly on applications in North America (e.g., physical pretreatment for digestion). Sludge reduction performance reported in the literature ranges from around 20% for some of the physical lysis processes to 80% for combination biological processes and some chemical treatment processes.

Understanding the mechanisms and process parameters is important in evaluating the applicability of the technology for a particular waste stream and economic conditions. Wastewater or sludge conditions that may impact performance include the following:

- Floc structure
- Divalent cations
- Availability of nutrients, including sulfur for anaerobic digestion – this is usually more important for industrial wastewater
- MLSS and TS concentrations for wastewater and digestion applications, respectively
- Inert, suspended, and biodegradable fractions

- Protein, carbohydrate, and lipid fractions
- Toxicity potential – this is usually more important for industrial wastewater
- Substrate temperature
- Chloride concentrations
- Polymer dose for thickening WAS

The success of a technology will also depend on the use of appropriate operational parameters, such as adequate input energy for physical processes and the economics of heat, footprint, sludge disposal, existing infrastructure, equipment, chemical costs, and so forth.

CHAPTER 3.0

PRELIMINARY EVALUATION

3.1 Introduction

This chapter provides a preliminary assessment of the WAS reduction processes carried forward from the Literature Review, Task 2 of the Evaluation of Processes to Reduce Activated Sludge Solids Generation and Disposal. The objective of this task was to select at least one technology from each fundamental process mechanism – biological, chemical, and mechanical. The representative technologies identified from this selection were carried forward for consideration in the following tasks.

The outline of this chapter is as follows:

- Technology Description
- Information Received from Project Installations
- Technology Shortlist
- Multi-criteria Decision Analysis

Vendor information was obtained, and independent verification was conducted by the project team. Although not all of the individuals were available for interview, an attempt was made to contact the representative installations to obtain high-level information in terms of process operation and performance. The team augmented the information and data to provide verification of process performance.

From the information gathered and the evaluation criteria developed, a short list of technologies and processes was developed. Technical criteria were integrated into a matrix, and the technologies were rated based on the team's experience of the technology and the information obtained from the vendors and operations personnel at the representative facilities. In addition, criteria were developed for the multi-criteria analysis (MCA), which is an initial step in the ultimate technology evaluation tool.

3.2 Information Received from Vendors

A questionnaire was developed for the collection of information from the vendors and operations personnel at representative facilities of the WAS-reduction technologies. A copy of this questionnaire is included in Appendix C.

Responses to the questionnaire are summarized in Tables 3-1 and 3-2. Appendix C includes the questionnaire submitted to the vendors. Appendix D includes a comprehensive list of installations as of 2008.

Table 3-1. WAS Reduction Technology Installation List for Biological Treatment Processes.

Manufacturer	Process	In Operation	Design Flow (MLD)	Municipal/ Industrial	General Range of Impacts	No. of Operational Installations
Biological Treatment Processes						
ENEA	Sequenced Wastewater Treatment	1998	0.013	Municipal	0.2 kg TSS/kg COD; 0.09 kg VSS/kg COD	1
Purac	Sequenced Wastewater Treatment	2003		Municipal	Up to 39% > VSr	1
Shaw Environmental	Sequenced Wastewater Treatment	2007				1
Vivendi	Sequenced Wastewater Treatment			Synthetic	58% less WAS	Bench-scale
ANOX AB	Sequenced Wastewater Treatment			1 Municipal; 1 Industrial	14% to 80% less WAS	1 Operational
Kuraray	Biological cell lysis	2001	0.02–6.8		70–100%	3 Industrial; 1 Municipal Large Bench-scale
PWI	Enhanced Solids Reduction with Multi-Action Conditioner					
United Utilities/ Monsal	Enhanced/ Enzymatic Hydrolysis	2002	12–54	Municipal	22% > VSr	4 Operational, 5 Construction
Siemens Cannibal®	Combination processes	1999	0.3–60.6	1 Industrial (Cheese Factory)		17 Operational; 11 Construction
CATABOL	Combination processes					1 Operational, 1 Pilot, 1 Construction
Aquarius Technologies	Combination processes		0.15–0.8	1 Industrial; 2 Municipal	No Biological Sludge	3 Operational

Notes:
COD = chemical oxygen demand
kg = kilogram(s)
MLD = million liter(s) per day
PWI = Premier Wastewater International
TSS = total suspended solids
VSr = volatile solids reduction
VSS = volatile suspended solids
WAS = waste-activated sludge

Table 3-2. WAS Reduction Technology Installation List for Chemical and Physical Treatment Processes.

Manufacturer	Process	In Operation	Design Flow (MLD)	Municipal/ Industrial	General Range of Impacts	No. of Operational Installations
Chemical Treatment Processes						
Praxair®	Ozonation	May 2006	26.5	Municipal with 60% Textile	35–80%	1 Operational
Degremont	Ozonation	4–12 mths				3 Pilot
University – Horng	UASB + Chemical Hydrolysis			Industrial – Maleic Dehydrate	90% less WAS	1 Operational
Heinzman and Engel	Thermal + Chemical Hydrolysis			Municipal		3 Operational
PMC Biotec	Chem/Thermophilic Digestion	Before 2000		Industrial vs. High Strength, to Sewer	>80%	5
Physical Treatment Processes						
Ultrawaves	Sonication	May 2003	1.4–110.14	Municipal WWTPs	15–35% >VSr and 0–10% >dewatering; 15–60% less WAS	20 Operational; 17 Pilot
Sonico	Sonication	2002	18–530	Municipal	VSr/WAS redn >20 to 30%; Dewaterability 0 to 3 percentage points	7 Operational; 7 Pilot
VTA	Sonication	Oct. 2003	4.2–36	Municipal	Up to 25% >VSr	9 Operational; 1 Pilot
CAMBI™	CAMBI™ Thermal Hydrolysis	1996		1 Industrial; 1 Combined	VSr >40–50%; dewaterability > 6 to 11% points	10 Operational; 1 Pilot; 6 Construction
Veolia	BioThelys thermal hydrolysis	2006	4.6–7.2	Municipal		2 Operational
MicroSludge®	Homogenization	2004 Pilots	17.5–27		35–90% > VSr	3 Pilots
Crown®	Pressure-release Disintegration	2002	2.16–102			16 Operational; 2 Construction; 6 Pilot
OpenCel	Pulsed Electric Field			Municipal	24% less WAS	1 Pilot Operational, 1 Pilot Completed
Kady®	Milling				12–50% less WAS	0 Operating, 4 Pilot, 1 Laboratory
Lysatec GMBh	Lysate centrifuge			Municipal	15 –26% > VSr	4 Operating

Notes:
MLD = million liter(s) per day
UASB = upflow anaerobic sludge blankets
VSr = volatile solids reduction
WAS = waste-activated sludge
WWTP = wastewater treatment plant

3.3 Technology Shortlist

This section includes the development of the criteria used for selection of the WAS-reduction technologies that were used for further investigation. Criteria were integrated into a matrix, and the technologies were rated based on the information gathered from the vendors, the information gathered from the facility questionnaires, and the team's experience. The objective was to develop criteria that would group the technology in the categories that would best facilitate achieving the overarching objectives of this WERF study, which is the development of a methodology and accompanying tool to facilitate the evaluation of WAS-reduction technologies. For the selected processes, laboratory analyses were conducted to further the understanding of the mechanisms for sludge reduction and particular sludge characteristics that contribute to the variability in performance. Hence, it was important that the technologies selected were representative of one of the following three "mechanistic" categories: biological, chemical, or physical. The criteria for technology grouping are as follows:

- Representative of other technologies within category
- Number of operating/full-scale installations
- Length of time technology has been in operation
- Availability of data
- Technology suitable for industrial wastewater
- Technology suitable for municipal wastewater
- Ease of retrofit

Each process was rated on a scale from 0 to 10 based on the previously listed criteria. Table 3-3 presents the scores given to each technology based on these criteria.

Table 3-3. Multi-Criteria Analysis for WAS-Reduction Technologies.

WAS-Reduction Technology	Representative of other technologies within category	Number of operating/full-scale installations	Length of time technology has been in operation	Availability of data	Technology suitable for industrial wastewater	Technology suitable for municipal wastewater	Ease of retrofit	Total
Biological								
Anaerobic Treatment of Wastewater	3	1	9	1	9	3	2	28
Anaerobic-Aerobic Sequenced Treatment	7	1	1	1	5	7	5	27
Predation Processes	1	1	8	1	3	2	2	18
Biological Cell Lysis	3	6	6	5	6	8	5	39
Combination Processes	**10**	**8**	**8**	**6**	**7**	**9**	**5**	**53**

Table 3-3. Multi-Criteria Analysis for WAS-Reduction Technologies, continued.

WAS-Reduction Technology	Representative of other technologies within category	Number of operating/full-scale installations	Length of time technology has been in operation	Availability of data	Technology suitable for industrial wastewater	Technology suitable for municipal wastewater	Ease of retrofit	Total
Chemical								
Ozonation	**8**	**1**	**10**	**7**	**6**	**5**	**7**	**44**
pH Adjustment	9	4	6	4	7	3	5	38
Physical								
Sonication	7	9	6	3	5	4	7	41
Thermal Hydrolysis	**5**	**7**	**10**	**6**	**6**	**7**	**5**	**46**
Homogenization – MicroSludge®	3	1	0	8	5	5	5	27
Pressure Release	**7**	**8**	**5**	**6**	**7**	**8**	**8**	**49**
Pulsed Electric Field	7	1	0	6	5	5	7	31
Shearing – ABS-Kady®	7	0	0	1	5	5	7	25
Lysate Centrifuge	6	3	9	1	5	6	7	37

Note:
WAS = waste-activated sludge

Table 3-4 presents the technologies that were chosen, the mechanistic principle category they represent, and the associated vendor. The physical processes were quite diverse, and so it was decided to include two physical processes for further consideration.

Table 3-4. List of WAS-Reduction Technologies Chosen for Further Analysis.

Category	Technology	Vendor
Biological	Combination Processes	Cannibal®
Chemical	Ozonation	Praxair®
Physical	Thermal Hydrolysis	Cambi™
Physical	Pressure Release	Crown®

CHAPTER 4.0

Collection and Evaluation of Laboratory and Field Data

4.1 Introduction

As stated previously, WERF seeks to establish a comprehensive evaluation methodology for WAS-reduction processes based on the in-depth consideration of a select number of technologies considered to be representative of the many options currently available in the marketplace. These technologies cover both activated sludge process reconfigurations aimed at reducing the generation of residual solids from the liquid treatment process as well as those designed to precondition WAS to make it more amenable to a subsequent stabilization process such as anaerobic digestion.

Based on the evaluation conducted in Chapter 2.0, representative technologies were chosen for further evaluation and are presented in Table 4-1. These four representative technologies were intended to provide validation of the processes and allow the establishment of differences in the net present value, variability in process performance, operability, and maintenance issues associated with different WAS-reduction technologies.

Table 4-1. List of WAS-Reduction Technologies Chosen for Further Analysis.

Category	Technology	Vendor
Biological	Combination Processes	Cannibal®
Chemical	Ozonation	Praxair®
Physical	Thermal Hydrolysis	Cambi™
Physical	Pressure Release	Crown®

The following list includes the facilities that provided samples for purposes of process validation to the lab at Virginia Tech. Although researchers gave sample kits and followed with numerous requests to the Praxair® facility, they received no samples. Hence, the discussion only considers Cannibal® (cyclic metabolic environment), Cambi™ (thermal hydrolysis), and Crown® (pressure-release).

- Cannibal® System:
  - Peru, Indiana
  - Emporia, Virginia
  - Big Bear, California
  - Morongo, California
- Cambi™ Process:
  - Naestved, Denmark

- DC Water and Sewer Authority pilot plant operated at Virginia Tech
- Crown® Process:
  - Rosedale WWTP, New Zealand

4.2 Discussion and Results

4.2.1 Cyclic Metabolic Environment – Cannibal®

The solubilization or release of organic matter in the interchange reactor of the Cannibal® process and subsequent degradation in the aerobic reactor has been proposed as a key indicator of achieving low sludge yields in the Cannibal® system (Novak et al., 2007). Therefore, it was thought that one possible method for assessing the solids reduction potential of the Cannibal® process was to measure the quantity of sCOD (or other measures of organic matter) released by incubating the sludge being fed to the interchange reactor under anaerobic conditions in a laboratory batch reactor. Data from such incubations conducted by Dr. John Novak at Virginia Tech are presented in Figure 4-1. This figure illustrates the results of anaerobic incubation tests, in which sCOD concentrations from sludge samples were measured over time in an effort to simulate the operation of the interchange reactor of the Cannibal® process.

The data for sCOD release were initially promising with the higher sCOD release being associated with a Cannibal® process with expected sludge yields and the lower sCOD release was from a Cannibal® process with sludge yields that were higher than expected. However, follow-up tests with other operating Cannibal® systems failed to confirm these findings. The sCOD release data were obtained for three other full-scale facilities that did not use the Cannibal® process. The three plants exhibited different sCOD release characteristics, suggesting that this assessment approach may have some merit as a prescreening methodology for the Cannibal® process; however, more research is needed to determine how effective this technique is as a prescreening method.

Similar data were found by Khanthongthip (2010) for laboratory studies conducted to determine the mechanisms that account for enhanced solids loss in the Cannibal® process. Khanthongthip found well-performing Cannibal® systems (Y = 0.12 g VSS/g COD), released 60% higher concentrations of sCOD over eight days compared with a poorly performing systems (Y = 0.32 gVSS/gCOD). In these tests, sludge was removed from the aeration basin of a laboratory system and allowed to batch incubate under anaerobic conditions for eight days.

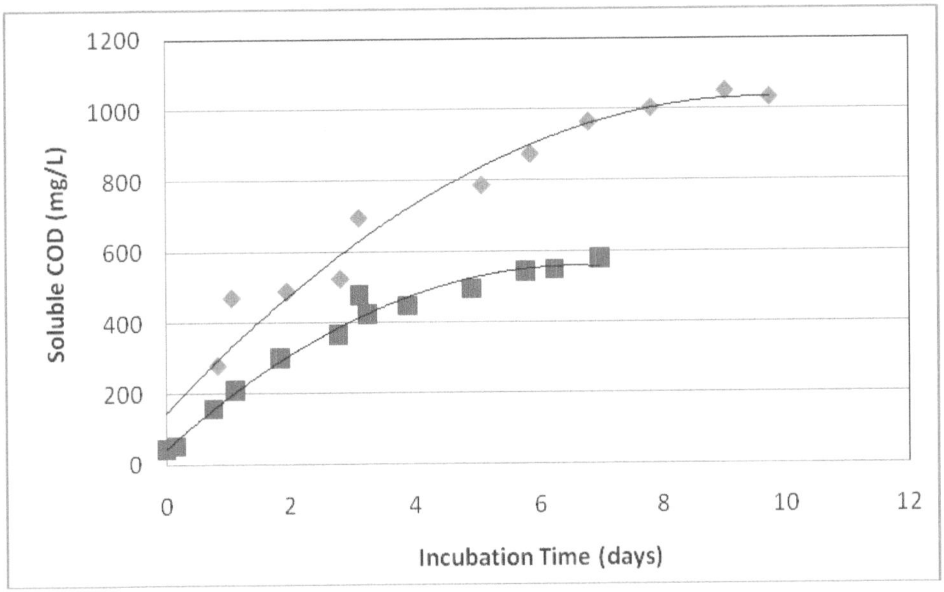

Figure 4-1. sCOD Data from two Cannibal® WWTPs Showing Differences in the Amount of COD Released over Time.

It appears that while the release of sCOD in the interchange reactor may be an important indicator of solids destruction performance by the Cannibal® process, a simple test to quantify the release over time under anaerobic conditions is not adequate to predict which plants will and which will not perform well with regard to the solids yield. Attempts to normalize the release data for volatile solids or total solids likewise did not improve the predictability of the sCOD release results. However, additional studies of this concept appear to be warranted.

4.2.2 Digestion Pretreatment Processes

4.2.2.1 Lab-Scale Evaluation

Pretreatment and digestion studies were carried out to evaluate the effects of pretreatment on the rate and extent of solids reduction. In addition, sCOD was measured, and for one of the sludges, electron micrographs were obtained following treatment. Both the solubilization of COD (and BOD) and the destruction of activated sludge flocs are indicators of the physical performance of pretreatment methods; however, these indicators alone are not indicative of improved digestion or solids reduction. Therefore, a goal of this portion of the project was to evaluate pretreatment technologies and compare the results with the digestion rate and extent of solids destruction.

The lab-scale study included sonication and shearing technologies, as these were the technologies available to the investigators. These physical pretreatment processes intend on lysing cells and breaking floc, similar to the full-scale technologies analyzed, which include thermal hydrolysis and pressure-release.

Methods. Sludges were obtained from three treatment plants. Two were collected near Virginia Tech and transported to the lab in containers. The other sludge was collected and shipped to Virginia Tech in coolers that contained ice to reduce biological activity.

For the two local sludges (called Sludge A and Sludge B), a mix of 90 milliliters of digested sludge from the plants and 360 milliliters of combined primary and secondary sludge (50%/50% by weight) was mixed and placed in 500-milliliter flasks and allowed to anaerobically digest for 57 days. The secondary sludge, 190 milliliters of WAS, was pretreated in the following manner:

- 10-minute shearing with a Waring blender at 8,000 to 10,000 revolutions per minute (rpm), 10-minute cooling in water-jacketed beaker
- Ozone dose of approximately 0.13 milligram O_3 per milligram TS using a Xetin Air Zone XT301 ozone generator with a generation rate of 300 milligrams O_3 per hour
- Untreated control

All reactors were completely mixed and maintained at approximately 37°C in an incubator. For this evaluation, gas was vented through water traps but not measured. Sampling occurred on days 0, 7, 14, 21, and 57. The COD analysis was conducted photometrically using Hach High-Range COD tubes (Hach Company, Loveland, Colorado), which had a range of 20 to 1,500 mg/L COD and are US EPA-approved. Initial and final total and volatile solids were measured according to Standard Methods (A.P.H.A. et al., 1995). pH was monitored with an Accumet® pH probe and meter.

After the studies were completed using the two local sludges (Sludges A and B), a third sludge was obtained. For this study, the activated sludge was treated in the following manner for various tests:

- 10-minute shearing with a Waring blender at 8,000 to 10,000 rpm, 10-minute cooling in water-jacketed beaker
- 5-minute sonication at 20 kHz, 166 watts, and sonication density of 0.33 W/mL using a Dukane DPC I 2120 generator and a Dukane 41C27 20-kHz mounted probe with a tip diameter of 37.5 millimeters and no additional focusing horn, 5-minute cooling in water-jacketed beaker
- Ozone dose of approximately 0.13 milligrams O_3 per milligram TS using a Xetin Air Zone XT301 ozone generator with a generation rate of 300 milligrams O_3 per hour
- Untreated control

The reactors were 1-liter Corning glass bottles with approximately 800 milliliters of sludge blended with a ratio of 1:1.5:1.5 digested:primary:WAS. The digested sludge was used as a seed sludge.

The results for Sludges A and B, Total and VSr data are shown in Figures 4-2 and 4-3. For Sludge A, no difference was found between the control and the samples receiving pretreatment. The data represent duplicate measures of the volatile solids and TS over time. For Sludge B, the data are not as consistent, with the ozone data showing unreliable results. This was primarily due to a large variation in the duplicate samples. These data indicate that for the sludges tested, no benefit was found to ozonation or mechanical shear for TS or VSr.

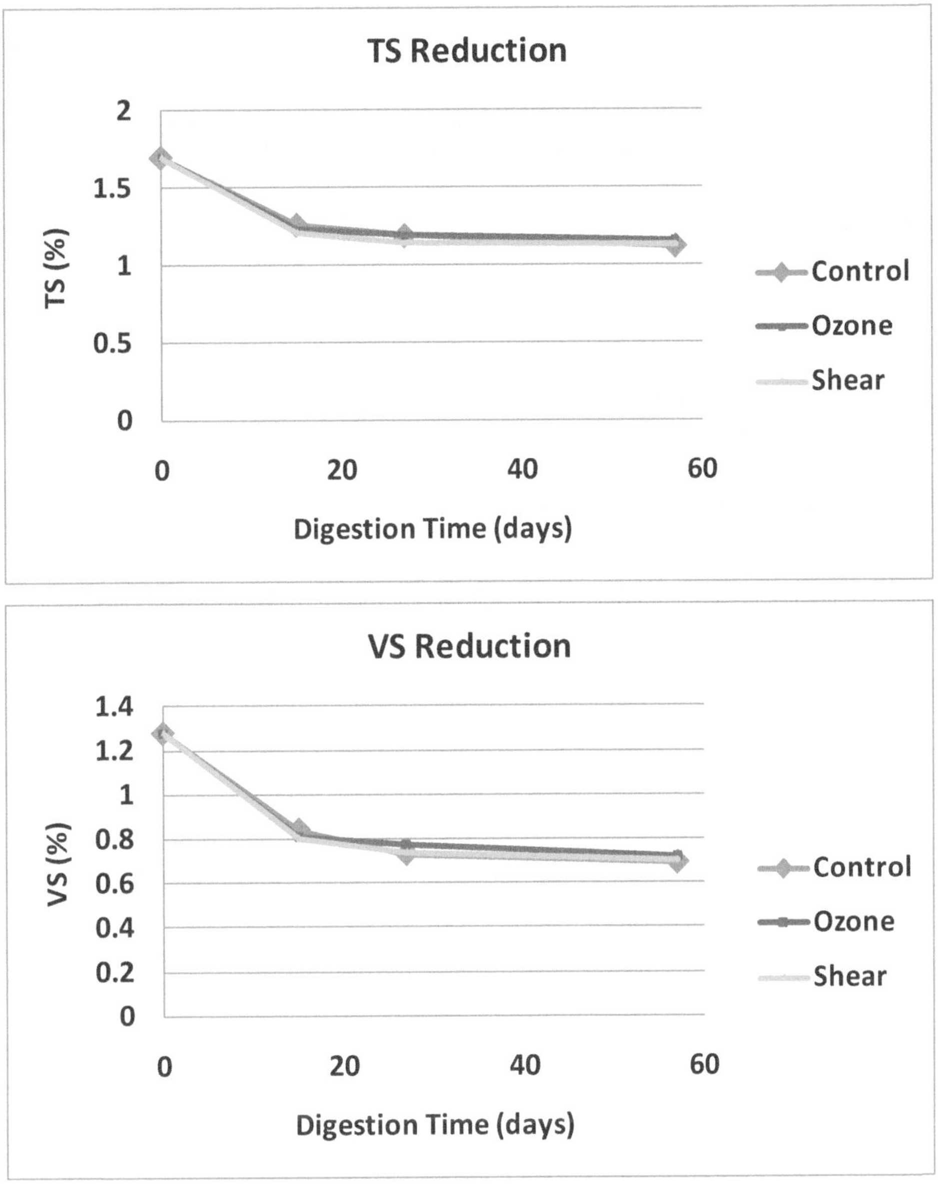

Figure 4-2. Effect of Pretreatment on TS and VSr for Sludge A.

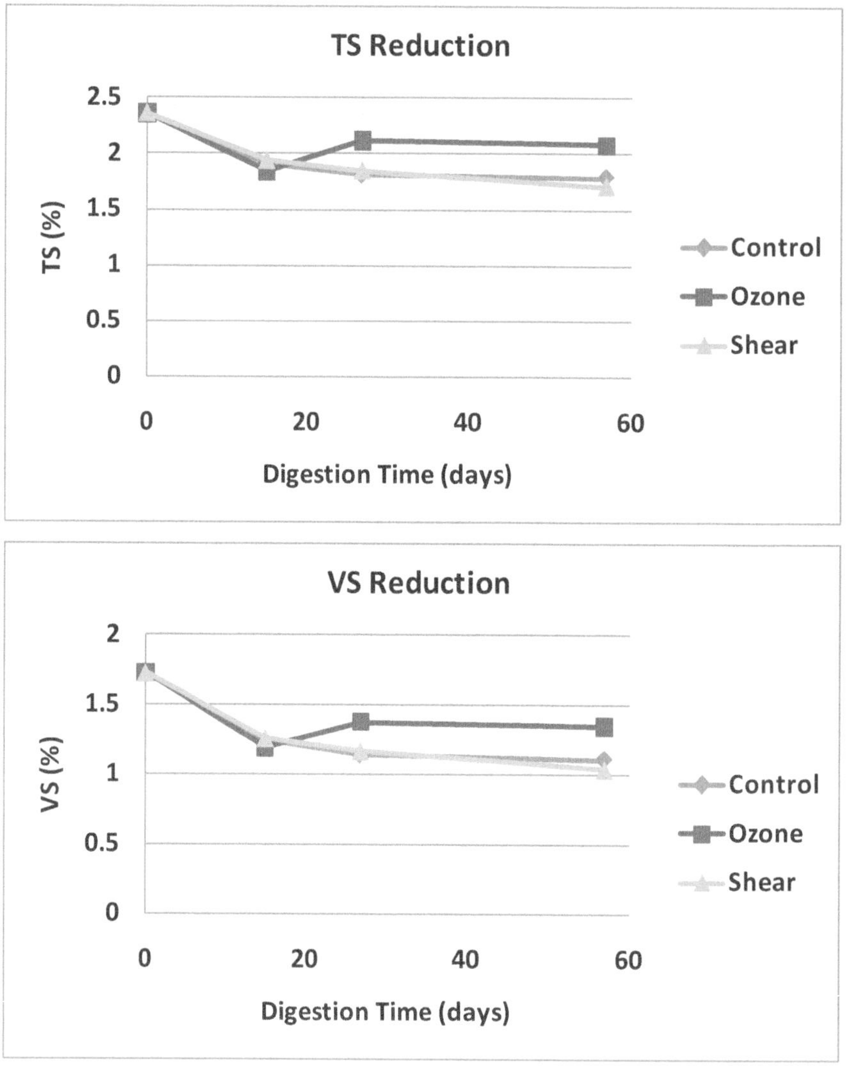

Figure 4-3. Effect of Pretreatment on TS and VSr for Sludge B.

Based on these results, a follow-up study was conducted using sludge from a utility in Ohio (called Sludge C). The utility was interested in determining if pretreatment would be beneficial for their digestion system. In addition to ozonation and mechanical shear, sonication was added to the pretreatment steps that were evaluated.

Data for TS, volatile solids, and COD for this sludge are shown in Figures 4-4, 4-5, and 4-6. The TS and VSr show little difference between the control and any of the pretreatment methods. The most notable effect of pretreatment is for the sludge that was sheared prior to anaerobic digestion. Shear appeared to slow the initial rate of degradation. However, by the end of three weeks, the degradation is similar to the control and the other pretreatment methods. The reduction in the degradation rate for the sheared sample could be due to the mechanical separation of hydrogen-producing organisms for hydrogen-consuming organisms. McCarty and Smith (1986) showed that interspecies transfer of hydrogen could influence the rate of digestion; so when mixing was excessive, the rate of volatile fatty use slowed. For this project, this effect is thought to be primarily a batch digestion effect. If the sheared sludge is being fed to a complete mix digester rather than being a part of the feed to a batch digester, the effects of mechanical shear on species separation would not be expected to slow the degradation rate.

Total COD data are shown in Figure 4-5. Little difference in the decline in COD can be seen for the various pretreatment methods. However, it can be seen that the lowest COD and lowest TS at the end of the degradation period is for the Control.

The sCOD is shown in Figure 4-6. The effect of pretreatment on sCOD can be clearly seen at the beginning of the degradation period. The lowest sCOD, as expected, is for the Control. The highest is for the ozone treatment, and the next highest is for the mechanically sheared sludge. It appears that the generation of sCOD using one of the pretreatment methods does not result in better overall degradation. The sCOD data also show that one effect of the microbial process – with or without pretreatment – is to generate sCOD so the physical or chemical methods to increase sCOD have little overall benefit for degradation. Another way of expressing this effect is that these pretreatment methods do not appear to convert non-biodegradable or slowly degradable organic matter to rapidly degradable organics.

In Figure 4-7, electron micrographs of the activated sludge following pretreatment are shown. It is clear from the electron micrographs that all the pretreatment methods appear to destroy or disperse biological floc. However, this floc dispersal does not result in increased degradation.

These laboratory studies show that some of the methods used to indicate the efficacy of pretreatment methods – specifically that these methods will increase sCOD and will destroy floc structures – cannot be used as indicators that pretreatment will increase solids destruction. Further, it appears that many of these pretreatment methods, when used at low to moderate application doses, will not improve degradation. On the other hand, it is known that very high doses of ozone can directly destroy organic matter, and a number of studies on mechanical shear and sonication show that increased solids destruction can occur.

Therefore, a complete assessment of the benefits of pretreatment technologies for a specific treatment plant appear to require pilot-scale or extensive laboratory studies. Just because a pretreatment process increases the soluble BOD or sCOD and destroys the floc structures does not indicate that the process will result in additional solids destruction.

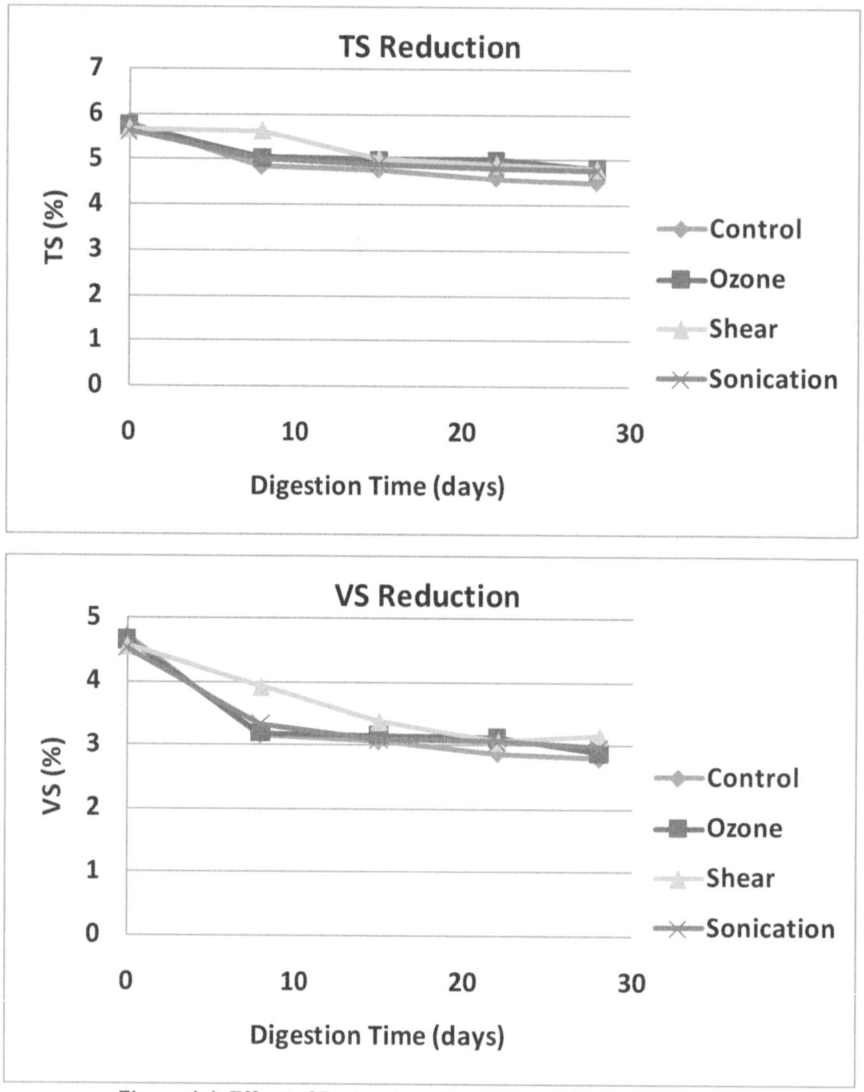

Figure 4-4. Effect of Pretreatment on TS and VSr for Sludge C.

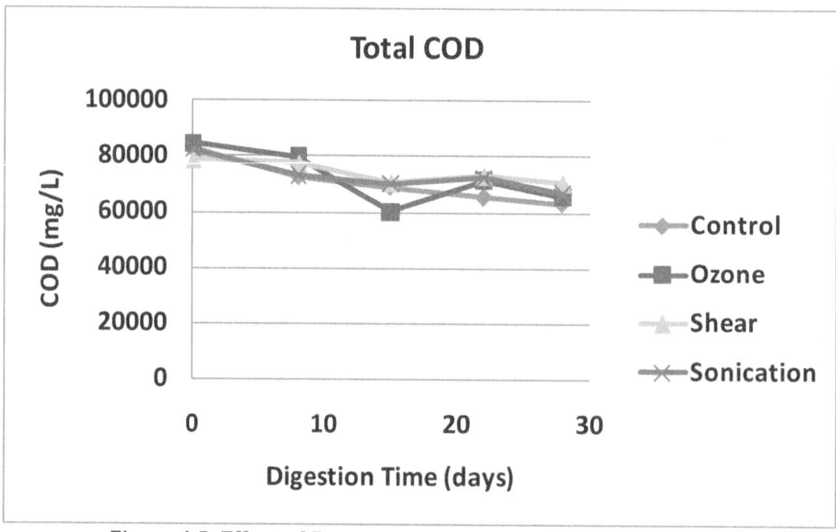

Figure 4-5. Effect of Pretreatment on Total CODr for Sludge C.

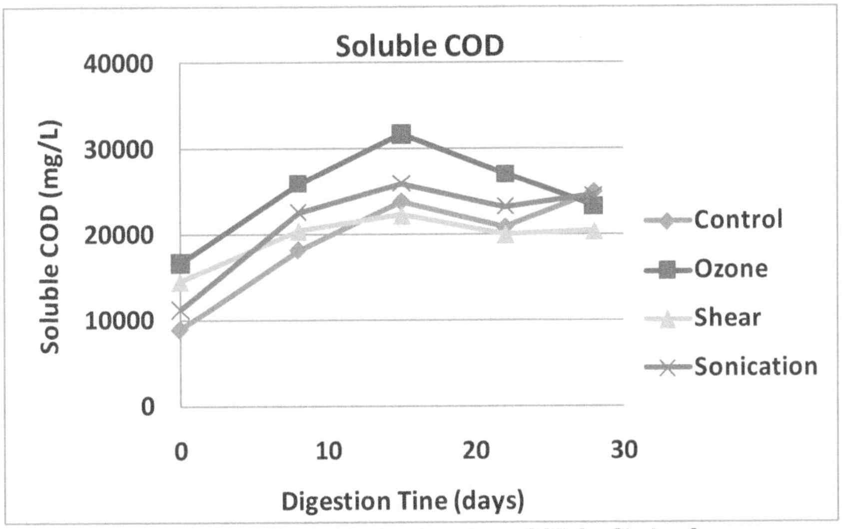

Figure 4-6. Effect of Pretreatment on sCOD for Sludge C.

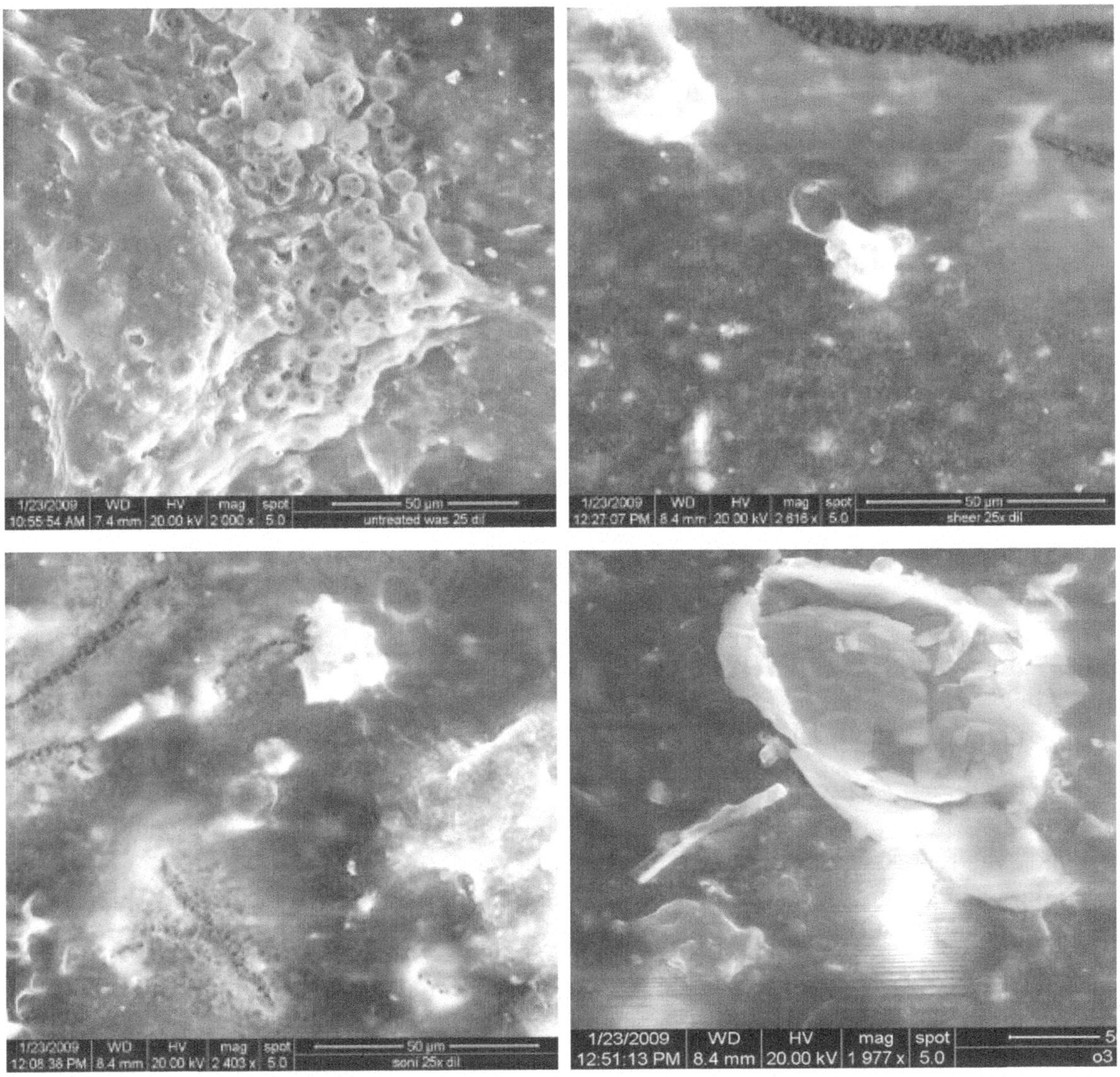

Figure 4-1. Upper Left – Untreated; Upper Right – Mechanically Sheared; Lower Left – Sonicated; Lower Right – Ozonated.

4.2.2.2 Full-Scale Evaluation

Thermal Hydrolysis – Cambi™. Data from thermal hydrolysis research recently conducted by Dr. John Novak at Virginia Tech for DC Water and Sewer Authority (DCWASA) were also evaluated as part of this project. During this pilot testing effort, VSr, ammonia concentration, and dewatering characteristics were investigated. Testing was performed using a control conventional mesophilic digester fed untreated sludge, and two digestion units fed thermal hydrolysis pretreated sludge. The digesters were operated at 15- and 20-day SRTs.

Table 4-2 summarizes the results of this data. Volatile solids reduction improved by 15 to 19%, and significantly higher ammonia concentrations were recorded from the Cambi™ pretreated units. Ongoing research by Novak links this increase in ammonia levels to the hydrolysis of proteins from the activated sludge biomass in the IR unit. It should be noted that the Cambi™ digester feed was 10% TS and the conventional was 5% TS – so digester loading was double or more than conventional.

Table 4-2. DCWASA Performance Data Summary for the Cambi™ Pilot Study.

Parameter	Conventional Mesophilic Digestion (20-day SRT)	Cambi™ – 15-day SRT	Cambi™ – 20-day SRT
VSr	50%	58%	60%
Ammonia	1,430 mg/L	2,446 mg/L	2,134 mg/L
Dewatered % DS	24%	34–36%	34–36%
Cake Odor*	N/A	Improved by 80–90%	Improved by 80–90%

Notes:
DS = dewatered sludge
mg/L = milligram(s) per liter
SRT = solids retention time
VSr = volatile solids reduction
* Measured as concentration of volatile organic sulfur compounds

The DCWASA pilot testing performance indicated that there was not a significant difference in solids reduction between the control and the Cambi™ units initially. However, after a year of operation, the solids destruction from Cambi™ increased by 8 to 10%. This is believed to be the result of biota initially being inhibited by the high ammonia concentrations resulting from the Cambi™ process and its eventual acclimation over time to these high levels. This suggests that short-term testing of a Cambi™ process with un-acclimated sludge may not provide data that will be consistent with full-scale operation.

Testing the impact of Cambi™ on sludge dewatering characteristics was conducted using a simulated centrifuge process. Results indicated that significant higher (approximately a 37% increase) dewatered cake concentrations could be achieved by this process. Reducing water content could result in significant savings for a treatment facility in terms of the hauling costs associated with offsite hauling of dewatered cake.

Pressure Release – Crown®. Tables 4-3 and 4-4 summarize data analyzed from several Crown® installations. According to this data summary, VSr improved on average by 28% and the TS concentration of dewatered cake improved by an average of 18%. No significant increase in biogas yield was observed.

Table 4-3. Crown® Performance Data Summary.

Site Name	PE	Location	Vol Trtd (m^3/hour)	VSr % Before	VSr % After	VSr % % Increase	Biogas production ft^3/lb VS des Before	Biogas production ft^3/lb VS des After	Biogas production ft^3/lb VS des % Increase
Wiesbaden Biebrich	130,000	Germany	12	32%	38%	20.0%	25.1	24.7	-1.7%
Taunusstein	30,000	Germany	6	32%	44%	38.9%	22.6	20.8	-7.8%
Ingelheim	200,000	Germany	6	36%	49%	34.1%	17.0	17.7	4.4%
Ginsheim	32,000	Germany	6	45%	54%	19.9%	14.7	14.3	-3.1%
Münchwilen	30,000	Germany	4	32%	43%	32.0%	20.2	19.1	-5.3%
Rosedale WWTP	650,000	New Zealand	6	51%	62%	21.6%	18.2	17.9	-1.8%
Average			**7**	**38.1%**	**48.3%**	**27.7%**	**19.6**	**19.1**	**-2.6%**

NOTES:
ft^3/lb = cubic feet per pound
m^3 = cubic meters
VS = volatile solids
VSr = volatile solids reduction
WWTP = wastewater treatment plant
Analysis of data not by WERF.

Table 4-4. Crown® Dewatering Performance Data Summary.

Site Name	DS after Dewatering % Before	DS after Dewatering % After	DS after Dewatering % % Increase
Wiesbaden Biebrich	31	36	16.1%
Taunusstein	31	36	16.1%
Ingelheim	28	34	21.4%
Ginsheim	20	23.4	17.0%
Münchwilen	22	26.4	20.0%
Rosedale WWTP	18.5	22.2	20.0%
Average	**25.1**	**29.7**	**18.4%**

NOTES:
DS = dewatered sludge
WWTP = wastewater treatment plant
Analysis of data not by WERF.

Sludge Homogenization – MicroSludge®. Another component of this study consisted in the independent evaluation of the data collected from the demonstration project conducted by Paradigm Environmental Technologies and the City of Des Moines using the MicroSludge® process, an anaerobic pretreatment technology that relies on both chemical and physical mechanistic principles. A description of this process is included in Chapter 1.0 of this report. A separate report presents a detailed description of the demonstration study.

The Des Moines, Iowa, Metropolitan Wastewater Reclamation Authority's WRF services approximately 400,000 population equivalent and generates about 5,500 dry tons of WAS annually. MicroSludge® was installed to evaluate its effect on biogas production and on the quantity of digested sludge residuals for disposal. At the WRF, MicroSludge® processing of TWAS followed by mesophilic co-digestion with PS and trucked liquid waste (TLW) was tested at full scale. Co-digestion of TWAS with PS at different digester HRTs and digestion of TWAS only at a constant HRT of 15 days were also tested at a pilot scale.

The demonstration study conducted at the Des Moines WRF provided an excellent opportunity to assess the performance of this WAS conditioning technology as applied at a full-scale installation. Although the tests conducted using full-scale anaerobic digesters were difficult to control, making it difficult to determine potential benefits from pretreating TWAS before digestion by this technology, it offered the opportunity to evaluate the operational reliability and maintenance requirements of a commercial-scale MicroSludge® system. The subsequent pilot-scale digester tests (which relied on feed from the full-scale MicroSludge® process) allowed the actual determination of process performance differences under much more controlled conditions. Following are the main conclusions that can be derived from this study:

Co-Digested PS and TWAS at Varying Digester HRTs

- VSr and CODr results appear to indicate that MicroSludge® technology may not be of much advantage to anaerobic digesters operating at or above 20 days HRT, and that its potential to increase the extent of sludge reduction is only achievable at lower HRTs. This result is consistent with previous studies at other facilities and suggests that MicroSludge® is capable of increasing the rate of digestion and, therefore, may reduce the retention time (thus digester volume) required for anaerobic digestion. Also, higher levels of VSr reduction at the lower digester HRTs would result in reduced quantities of digested biosolids, requiring further in-plant processing and ultimate disposal or reuse outside the treatment facility.
- Cumulative gas production was higher in the test unit than in the control under all the evaluated digester HRTs, even though results also showed that at a 20-day HRT there was no significant difference in VSr and CODr levels. An increase in biogas yield at essentially equivalent VSr and CODr levels could be explained by a difference in the characteristics of the methanogenesis substrate, as supported by the significantly higher VFA concentrations and higher methanogenic activity of the MicroSludge® pretreated sludge. Higher biogas production capabilities would increase the co-generation potential at a treatment facility.
- Results indicate a definite trend of increased biogas production rate as the HRT was decreased in both the control and test digesters, with even higher rates for the MicroSludge®-treated TWAS. Again, this finding would support the use of shorter detention time anaerobic digesters (~10- to 15-day HRT) with significant capital and O&M cost savings as compared with units sized for 20-day HRT or above.
- Organic sulfur emissions from MicroSludge®-treated sludge were about one-third of that of the control digester, indicating lower odor generation potential for MicroSludge®-treated sludge.
- Anaerobic incubation batch testing incorporating bromoethane sulfonic acid addition identified higher methanogenic activity for MicroSludge®-treated TWAS.
- Batch incubation testing also indicated that MicroSludge® pretreatment can lead to the direct formation of VFAs due to solubilization of sludge samples. The presence of significantly higher amounts of VFAs will consequently reduce the reaction time required for anaerobic digestion, since hydrolysis and acidogenisis (the necessary preceding steps before

methanogenesis) have already occurred. By comparison, an acid-phase anaerobic digestion system is typically designed for a 2-day SRT; hence, these data suggest that the rate of digestion is significantly increased by the MicroSludge® process.

- There were no statistically significant differences between the bacterial densities of the test and control digesters.

TWAS Only at 15 Days Digester HRT

- Tests indicated a much larger difference in the VSr between control and test digesters fed TWAS only when compared with the results from co-digestion at the same 15-day digester HRT. This could be explained by the significant presence of volatile solids in the PS that are not affected by MicroSludge® treatment. On the other hand, the values of CODr observed at Des Moines were very similar to those observed in the co-digestion tests. This could be explained by understanding that the basic mechanism behind MicroSludge® treatment is the lysing of cell mass, which is relatively difficult-to-degrade suspended matter, the subsequent production of high levels of sCOD (mainly VFAs), and from the fact that the organic matter in PS is largely unaffected by this sludge conditioning process.
- Biogas data showed a significant difference between control and test digesters total biogas production, which is considerably greater than the values reported for VSr and CODr. This suggests (similarly to de co-digestion test results) that the biogas yield may be improved by the MicroSludge® pretreatment of the TWAS. Again, these results would indicate that incorporating a MicroSludge® process ahead of an mesophilic anaerobic digester for TWAS would allow the use of shorter HRTs (thus reducing facility requirements); it would produce higher biogas quantities (higher co-generation potential), as well reducing the quantities of digested sludge and thus lowering the requirements for biosolids processing and disposal or reuse when compared against digestion of untreated TWAS only.
- Microscopic evaluations conducted indicate that MicroSludge® processing of TWAS destroyed filamentous organisms and thus would result in lower filamentous growth-related foaming incidents in anaerobic digesters.
- Sulfur-related emissions generated from both untreated and MicroSludge® pretreated TWAS samples were observed to be similar and generally very low compared with the results obtained with co-digestion of sludges. This indicates that the primary source of sulfur-related odors at Des Moines WRF was PS, and not the TWAS. Batch digestion tests also indicated very little difference in the methanogenic activity of the anaerobically incubated TWAS samples. This is in sharp contrast with the results from the co-digestion tests, in which significantly higher methanogenic activity was reported from samples of the MicroSludge®-treated digester. This difference could be explained if the PS is considered to be the source of both sulfur-related odor precursors and sulfur-reducing bacteria. It should be noted that there are sources other than sulfur that would result in noxious odor releases as part of the anaerobic digestion process and that were not evaluated as part of this study.
- Results at Des Moines showed less than a log reduction in pathogenic indicator organism as a result of treatment of the TWAS samples with MicroSludge®. Based on these results, it can be concluded that MicroSludge® treatment did not have a significant impact on pathogen reduction ahead of the anaerobic digestion process.

CHAPTER 5.0

DEVELOPMENT AND DEMONSTRATION OF EVALUATION TOOLS

5.1 Introduction

A suite of tools was designed as part of the WAS-reduction technology evaluation methodology developed under this project and includes a facility sampling and laboratory test protocol to determine relevant sludge characteristics; a process simulation approach to facilitate the simulation for a given facility; and an MCA and life-cycle cost module for decision making that allows the consideration of technical, financial, and nonfinancial criteria.

5.2 Multi-Criteria Analysis Tool (MCA)

The objective of the MCA Tool is to enable an analysis of the variables involving different WAS-reduction technologies. Multi-criteria analysis is a business decision-making method that is typically applied in situations involving selection among alternatives. It integrates what one can do (alternatives) with what one knows (data and assumptions) and does not know (uncertainties), and with what decision makers want (values). The process provides support to differentiate between options or decisions for an optimal strategy to resolve the targeted issue. The process is used to focus discussions on the elements of the problem (values, assumptions, or uncertainties) that actually affect the decision, thereby eliminating irrelevant, confusing, and conflicting issues. This allows consideration of decision makers' key concerns and assumptions and facilitates achieving consensus when stakeholders hold different opinions.

5.3 Facility Sampling and Sludge Testing Protocol

The objective of creating a facility sampling and sludge testing protocol is to develop a set of sludge characteristics that can be used to assess the impact of sludge-reduction processes for a particular wastewater treatment facility. Sludge and wastewater samples were collected from several installations as part of this study, and that experience, along with the understanding of the basic mechanistic principles behind each of the technology alternatives, served as the basis for the development of a recommended protocol, which is included in Appendix A to this report, for assessing relevant sludge characteristics for any treatment facility.

5.4 Process Simulation Approach

This section discusses the general framework to simulate WAS-reduction technologies and presents a recommended general process modeling approach for them. As mentioned in previous sections, several technologies were selected as representative of general mechanistic principles under which most of WAS-reduction technologies can be grouped. In some cases, data from actual facilities was used in the development of the modeling approach. In other cases (as in sludge ozonation), the lack of actual data did not preclude the development of a modeling approach, as it was possible to develop one relying primarily on the understanding of the basic chemical mechanistic principles behind the process.

The proposed process simulation approach defined herein relies on the use of nonproprietary models from the International Water Association (IWA): Activated Sludge Model (ASM) 2d for liquid stream biological process and Anaerobic Digestion Model No. 1 (ADM1) for anaerobic digestion (Batsone et al., 2002). The proposed process simulation approach can be calibrated to site-specific performance to assist in better understanding of a particular sludge minimization technology, evaluating its applicability for a particular situation, and establishing important process parameters.

5.4.1 Approach for Simulating Liquid Treatment Processes

5.4.1.1 Ozonation Process

The ozonation process has been applied either on a portion of the RAS stream, such as the Praxair® process, or the mixed liquor from aeration basins, such as Biolysis® "O" process. Both types are hypothesized to work through the same mechanisms (i.e., cell lysis or significant cell wall weakening through extreme oxidation), thus releasing cellular organic matter for further degradation. Cell lysis resulted from ozonation allows biota to metabolize sCOD within the conventional aerated treatment process, such as activated sludge. This, in turn, results in increasing OUR and reduced sludge yield. The modeling approach discussed here assumes that ozonation is being applied to a portion of the RAS stream.

Simulation of the ozonation process can be achieved by modifying the standard ASM 2d through incorporating the following assumptions:

1. CODr is achieved by oxidation and mineralization caused by ozonation. At the relatively low ozone doses used in this process, this reduction should be minimal. Only particulate COD fractions can be assumed to be reduced. This assumption can be further refined on the basis of actual bench-scale tests conducted on actual sludge samples.
2. Non-biodegradable particulate COD (Xi) is converted to slowly biodegradable particulate COD (Xs) as a result of ozonation.
3. Anoxic/aerobic biomass is converted to Xs and inert decay products according to inactive fraction of the biomass.
4. Xs is converted to soluble fermentable substrate (SF) (i.e., non-VFA portion of readily biodegradable COD [Ss] in the ASM 2d model).
5. Phosphorus-accumulating organisms (PAO) storage product (XPHA) is converted to Xs.

The extent of the conversions depends on ozonation dosage and specific RAS characterization. Due to lack of available research on effect of ozonation on various COD fractions, identical conversions can be assumed for all impacted COD fractions, until further study results are adequate to differentiate the effects between various COD fractions. Site-specific adjustment should be allowed for following parameters:

1. Percent of RAS ozonated
2. Percent COD reduction
3. Percent COD conversion

The nitrogen and phosphorus balances should be maintained before and after the ozonation by accounting for different nitrogen and phosphorus fractions of various COD components.

The proposed simulation of ozonation process was implemented for a hypothetic WWTP to assess the impact of the previously listed parameters. For simplicity, only the activated sludge process is modeled. Table 5-1 compares the sludge yield for this hypothetic WWTP with and without ozonation.

Table 5-1. Simulation of WAS Reduction by RAS Ozonation.

Simulated Scenarios	With Ozonation		Without Ozonation
Flow (MGD)		1.00	
Temperature (°C)		20	
Influent $CBOD_5$ (lb/day)		2,086	
Influent TSS (lb/day)		2,003	
% Influent VSS/TSS		75%	
Ammonia – N (lb/day)		250	
Total Phosphate – P (lb/day)		38	
SRT (days)		14.00	
Volume of Aeration Basins (MG)		0.50	
% RAS Ozonated	10%		0%
% COD Reduction by RAS Ozonation	5%		0%
% COD Conversion by RAS Ozonation	30%		0%
Simulation Results			
Average MLSS (mg/L)	2,833		3,743
Corresponding WAS (lb/day)	796		1,068
Biomass Yield Reduction (%)	24%		

Notes:

°C = degree(s) Celsius
$CBOD_5$ = carbonaceous biochemical oxygen demand 5-day test
COD = chemical oxygen demand
lb = pound(s)
MG = million gallons
MGD = million gallons per day
mg/L = milligram(s) per liter
MLSS = mixed-liquor suspended solids
RAS = return-activated sludge
TSS = total suspended solids
VSS = volatile suspended solids
WAS = waste-activated sludge

A COD reduction of 5% and COD conversion of 30% are used in the model, resulting 24% reduction in biomass yield. This is generally consistent with the operating data of WWTPs that have ozonation systems installed using relatively low-ozone dosages (Sandino et al., 2008).

5.4.1.2 Cannibal® Process

The Cannibal® process consists of both physical and biological components. The physical process uses very fine drum screens to remove Xi and hydrocyclones to remove grit and dense inorganic particles. The mechanism behind the biological process used by the Cannibal® process is not yet fully understood, but it is hypothesized that the decay products resulting from heterotrophs grown under aerobic and anoxic conditions become slowly degradable under anaerobic conditions established in the interchange reactors (Johnson et al., 2008). The ASM 2d can be readily modified to capture the physical and biological components of the Cannibal® process, by assuming the following:

1. Non-biodegradable COD and inorganic inert materials are removed physically from the RAS portion going through the fine screens and the hydrocyclones. The amount of Xi removal has been found to range from 90 to 100% in some WWTPs, but it can vary significantly from plant to plant.
2. The biological process of hydrolyzing the biological decay products can be modeled by including an additional state variable, Xdaa, and associated process rate equation, as well as modification of the stoichiometric variables of ASM 2d. The hydrolysis of Xdaa to Xs is similar to the hydrolysis of Xs. An identical hydrolysis rate constant is assumed for both hydrolysis processes (1.2 day-1), which will be subject to further refinement when more data are available (Johnson et al., 2008).

This simulation approach process was calibrated to historical data collected as part of this study from the Gill's Creek WWTP in South Carolina. The plant has operated with a Cannibal® process since March 2007. The interchange bioreactor is 1.5 million gallons (MG) in volume. The activated sludge process has an operating volume of 5.4 MG. It consists of two identical trains, each having four zones, two Verticel reactors followed by two fine bubble diffuser reactors. The Verticel reactors are equipped with coarse bubble diffusers and disc aerators to achieve mixing and aeration. The volume of the four zones in each train is 0.9 MG (Verticel), 0.9 MG (Verticel), 0.45 MG (with fine bubble diffusers), and 0.45 MG (with fine bubble diffusers), respectively. The ORP meters in the Verticel reactors typically read -175 millivolts (mV) in the first one and -150 mV in the second. There are no primary clarifiers.

According to Siemens, the WAS solids and screenings/grit removed from the physical components of the Cannibal® process (not including effluent solids) averaged at 0.1 to 0.2 pound TSS per pound BOD applied to bioreactors.

The available plant data from March to October 2007 were reviewed and used to calibrate the model. Major model parameters and outputs are presented in Table 5-2. As shown in the table, the simulated sludge yield matches the plant data well.

Table 5-2. Simulation of WAS Reduction by the Cannibal® Process.

Parameters	Plant Data	Calibrated Model
Influent Flow (MGD)	8.78	8.78
Influent $CBOD_5$ (lb/day)	11,767	11,767
Influent TSS (lb/day)	11,414	11,414
Influent VSS/TSS		75%
Influent Ammonia (lb/day) TKN	1,765	1,765
Temperature in the Biological Process (°C)	23	23
Removal of Inorganic Particulates across Grit Removal System		80%
Total Bioreactor Volume (MG)	6.30	6.30
Interchange Tank Volume (MG)	1.5	1.5

Table 5-2. Simulation of WAS Reduction by the Cannibal® Process. (continued)

Parameters	Plant Data	Calibrated Model
DO in Interchange Reactor (mg/L)		0.02
DO in Verticel 1 (mg/L)	(-175 mV ORP)	0.05
DO in Verticel 2 (mg/L)	(-150 mV ORP)	0.05
SRT in interchange tank (days)		10
SRT (days)		150
% of Influent Organic Xi Captured		50%
% of Influent Inorganic Inerts Captured		25%
Mixed Liquor Recycle (% of influent)	200%	200%
Secondary Clarifier Surface Area (square feet)	31,689	31,689
Effluent TSS (mg/L)	6	6
Simulation Results		
Corresponding Average MLSS (mg/L)	5,000	5,027
MLVSS (mg/L)	3,850	3,781
MLVSS/MLSS (%)	77%	75%
WAS TSS (mg/L)	10,696	11,683
WAS TSS (lb/day)	1,947	2,127
ScreeningGrit TSS (lbs/day)		927
Effluent TSS (lbs/day)		439
Observed Yield (lb WAS and Screening-grit TSS/lb BOD_5)	0.25-0.3	0.26

Notes:
°C = degree(s) Celsius
$CBOD_5$ = carbonaceous biochemical oxygen demand 5-day test
BOD_5 = biochemical oxygen demand 5-day test
DO = dissolved oxygen
lb = pound(s)
MG = million gallons
MGD = million gallons per day
mg/L = milligram(s) per liter
MLSS = mixed-liquor suspended solids
ML VSS = mixed-liquor volatile suspended solids
mV = millivolt(s)
ORP = oxidation-reduction potential
SRT = solids retention time
TKN = total Kjeldahl nitrogen
TSS = total suspended solids
VSS = volatile suspended solids
WAS = waste-activated sludge
Xi = non-biodegradable particulate chemical oxygen demand

5.4.2 Approach for Simulating Digester Pretreatment Processes

Although ADM1 can be used alone for anaerobic digestion process, it is often integrated with a whole plant simulator that uses a variation of IWA's ASM 2d for secondary treatment. The reason is that sufficient characterization of various digester feed sludge (e.g., PS and WAS), as required for ADM1, is usually not available. Coupling with a well-calibrated whole plant simulator can provide reasonable estimates of sludge characteristics to be used as ADM1 inputs.

Since ADM1 and ASM 2d were developed independently for fundamentally different anaerobic and aerobic biological processes, different sets of state variable are used. Conversions between the variable sets of ADM1 and ASM 2d are critical. Figure 5-1 presents the conversion paths for variable sets that are used for COD fractionations between ADM 1, Pro2D model, and ASM 2d (Johnson and Shang, 2005).

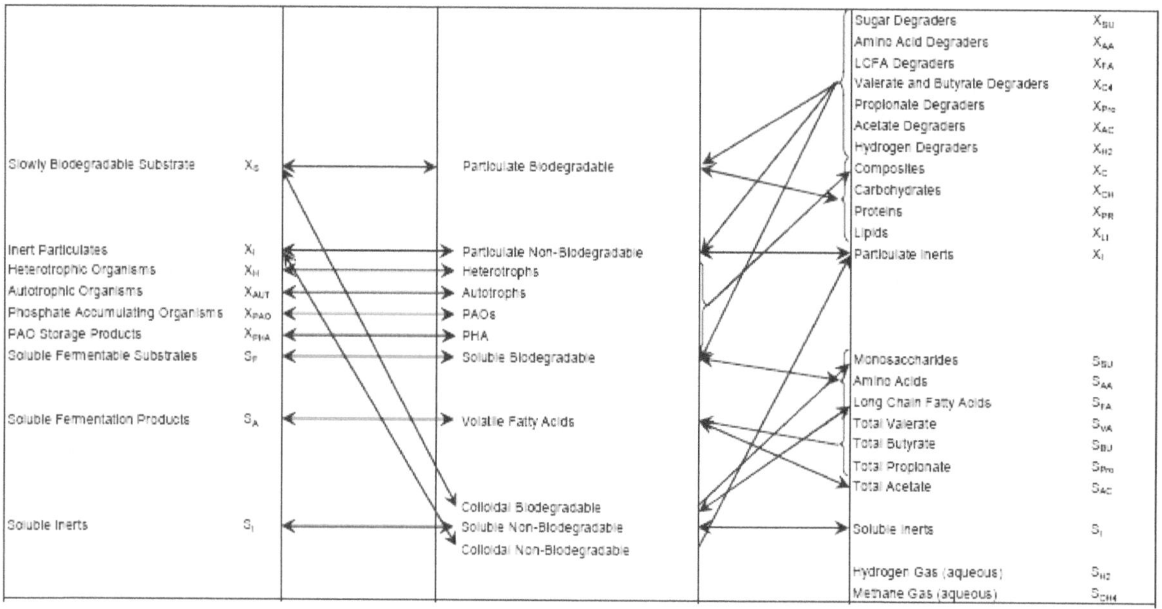

Figure 5-1. Conversion of COD Fractionations between ADM 1 and ASM 2d.

Waste-activated sludge solids are very different from PS from a digestability perspective. ADM1 defines anaerobic digestion as simultaneous biological processes of disintegration, hydrolysis, and acid and methane generations. It is reasonable to assume that at least some portion of PS will have gone through the disintegration step when it enters the digesters, as a result of prefermentation in the upstream sewer system and treatment processes. Therefore, ADM1 places the WAS (i.e., the active anoxic/aerobic biomass) in the composite fraction. It distributes the Xs fraction from the PS between the composite fraction and the proteins, carbohydrates, and lipids. For the purpose of illustration, the Xs fraction from the PS is placed entirely into the proteins, carbohydrates, and lipids.

Both of the thermal hydrolysis and pressure-release processes fall into the physical process group. All physical processes are aimed to induce floc disintegration and cell lysis to release organic material in the cells for biological consumption. Depending on the energy and intensity input of the pretreatment, a continuum of the following mechanisms are possible to realize in solids minimization:

1. Increasing disintegration rates (K_{dis}) of the various digester feed solids to carbohydrates, proteins, and lipids, as well as increasing hydrolysis rate (K_{hyd}) of carbohydrates (Xch), proteins (Xpr), and lipids (Xli) into respective monomers. This mechanism can be associated with all physical pretreatments since floc disintegration is typically resulted and a faster hydrolysis is possible due to more homogeneous environment and better contact between substrates and functional organisms.
2. Increased extent of the anoxic/aerobic biomass disintegration prior to entering digesters (i.e., essentially skipping the disintegration step). This is likely to occur for all pretreatment

processes since they can potentially break up cell membranes and release the organic matters inside.

3. Decay products of the anoxic/aerobic biomass (Xdaa) converted to carbohydrates, proteins, and lipids. The decay products are non-biodegradable under anoxic/aerobic conditions. Theoretically, they may be susceptible to biodegradation under anaerobic condition through a completely different pathway. However, keeping them non-biodegradable in ADM1 normally provides model predictions that are consistent with observed digestion performance. With pretreatment processes using chemical or thermal procedure (e.g., thermal hydrolysis process), the change of biodegradability is likely to be triggered through potentially induced new/modified degradation pathways.
4. Non-biodegradable particulate COD (Xi), representative of the non-biodegradable portion of PS, converted to carbohydrates, proteins, and lipids. Similar to anoxic/aerobic biomass decay products, non-biodegradable particulates are normally inert under anaerobic condition. A change of biodegradability may be achieved with pretreatment processes that use chemical or thermal processes (e.g., thermal hydrolysis process).

For Mechanisms 2 through 4, the distribution among carbohydrates, proteins, and lipids after the conversion can be determined by conducting nitrogen and phosphorus mass balances. Since the anoxic/aerobic biomass, its decay products (Xdaa), and non-biodegradable particulates (Xi) have their respective, associated nitrogen fraction, the ADM1 standard nitrogen fraction for proteins can be used to determine how much COD will be associated with the protein fraction. The remaining COD was then split between carbohydrates and lipids so that the phosphorus in the various feed solids is maintained in the phosphorus associated with proteins and liquids after the conversion.

To illustrate potential effects of these possible mechanisms, a hypothetic WWTP was simulated using default parameters of ASM 2d and ADM1. Then the previously discussed parameters are adjusted to demonstrate their impact on digester performance. The liquids treatment stream of the simulated plant includes primary clarification, activated sludge process, and tertiary filtration. Primary sludge and WAS are thickened and combined for anaerobic digestion and subsequent dewatering. Table 5-3 presents the digester performances responding to various parameter adjustments. Parameters that are different from standard ASM1 default values are shaded.

Table 5-3. Modeled Digester Performance with Cambi™ and Crown® Pretreatment.

Modeled Scenarios	With Pretreatment					No Pretreatment (ADM1 defaults)
Pretreatment Mechanisms	2,3,4	2, 3	3	2	1	
Applicable Pretreatment Processes	Cambi™	Cambi™	Cambi™	Crown®, Cambi™	Crown®, Cambi™	
Influent Flow (MGD)	1.0					
Temperature (°C)	20					
Influent $CBOD_5$ (lb/day)	2,086					
Influent TSS (lb/day)	2,003					
% Influent VSS/TSS	75%					
Ammonia –N (lb/day)	250					
Total Phosphate – P (lb/day)	38					
Primary TSS Removal Efficiency (%)	60%					
Aeration Basin SRT (days)	8.0					
Digester SRT (days)	20					
Parameters for Anaerobic Digestion						
% Xdaa Converted to Xch, Xpr, and Xli	**100%**	**100%**	**100%**	0%	0%	0%
% Anoxic/Aerobic Biomass Converted to Xch, Xpr, and Xli	**100%**	**100%**	0%	**100%**	0%	0%
% Non-biodegradable Particulate Converted to Xch, Xpr, and Xli	**100%**	0%	0%	0%	0%	0%
Hydrolysis Rate of Xch, Xpr, and Xli (1/day)	10	10	10	10	**10,000**	10
Disintegration Rate (1/day)	0.5	0.5	0.5	0.5	**1,000**	0.5
% Particulate Inert from Xc	0.25	0.25	0.25	0.25	0.25	0.25
% Soluble Inert from Xc	0.1	0.1	0.1	0.1	0.1	0.1
Modeled Results						
Primary Sludge (lb/day)	1,148	1,179	1,199	1,227	1,226	1,229
Average MLSS (mg/L)	2,524	2,520	2,517	2,543	2,541	2,536
Corresponding WAS (lb/day)	1,187	1,185	1,184	1,197	1,196	1,194
Combined Thickened Feed Sludge to Anaerobic Dig. (lb/day)	2,445	2,465	2,478	2,512	2,511	2,511

Table 5-3. Modeled Digester Performance with Cambi™ and Crown® Pretreatment, continued.

Modeled Scenarios	With Pretreatment					No Pretreatment (ADM1 defaults)
Pretreatment Mechanisms	**2,3,4**	**2, 3**	**3**	**2**	**1**	
Applicable Pretreatment Processes	**Cambi™**	**Cambi™**	**Cambi™**	**Crown®, Cambi™**	**Crown®, Cambi™**	
Model Prediction of VS Destruction (%)	88%	71%	61%	48%	48%	46%
Digester Gas Production (ft^3/day)	20,623	17,617	15,658	13,588	13,444	13,101
Anaerobic Digested Sludge (lb/day)	815	1,106	1,299	1,566	1,562	1,590
Increase in Digester Gas Production (%)	57%	34%	20%	4%	3%	
Increase in VS Destruction (%)	90%	54%	32%	3%	3%	

Notes:
°C = degree(s) Celsius
CBOD5 = carbonaceous biochemical oxygen demand 5-day test
ft^3 = cubic feet
lb = pound(s)
MGD = million gallons per day
mg/L = milligram(s) per liter
MLSS = mixed-liquor suspended solids
SRT = solids retention time
TSS = total suspended solids
VS = volatile solids
VSS = volatile suspended solids
WAS = waste-activated sludge
Xch = carbohydrates
Xc = composites (generic organic compound)
Xdaa = anoxic/aerobic biomass
Xli = lipids
Xpr = proteins

For Mechanism 1, the disintegration and hydrolysis rates are increased to several magnitudes higher. However, only minimal improvement is resulted in terms of VSS reduction and digester gas production. This can be explained by recognizing that methane generation, rather than disintegration and hydrolysis, is the limiting step of the anaerobic digestion. Thus increasing disintegration and hydrolysis rates will not extend the extent of the digestion. For a similar reason, Mechanism 2 alone (i.e., WAS skipping disintegration step) does not significantly improve the digestion either.

For Mechanisms 3 and 4, anoxic/aerobic biomass decay products (Xdaa) and non-biodegradable particulates (Xi) are assumed completely converted to carbohydrates, proteins, and lipids, respectively, to assess the maximum potential of individual mechanisms in impacting digestion performance. Significant increases in VSS reduction and digester gas production are predicted for both mechanisms. Mechanism 3 alone can increases the VSS reduction and digester gas production by up to 32 and 20%, respectively, because otherwise the decay products of anoxic/aerobic biomass would be inert, lowering the digestion performance. Similar effect is observed for Mechanism 4. With Mechanisms 2, 3, and 4 all combined, the VSS reduction and digester gas production can be increased by up to 90 and 57%, respectively.

The digestion performance improvement predicted for a combination of the discussed mechanisms covers the reported performance for Cambi™ and Crown® processes. The Cambi™

process has been reported to result in approximately 25% increase in gas production. A survey of Crown® processes found 20-39% increase in VSS reduction and 16- 40% increase in gas production. The exact extent of each mechanism in effect for a specific application has to be determined through site-specific testing.

The Crown® process does not involve chemical addition or a thermal step and is not expected to significantly achieve Mechanisms 3 and 4. However, its modeled improvement to the digestion process goes beyond the maximally achievable extent that is predicted for Mechanisms 1 and 2. Mechanisms 3 and 4 have to be in place to make the claimed improvement happen. Extensive studies/testing will be needed to better understand the underlying mechanisms and verify the existence of Mechanisms 3 and 4 for pretreatment such as the Crown® process.

CHAPTER 6.0

DISCUSSION AND NEXT STEPS

Sludge reduction mechanisms generally fall into the following three categories: 1) mechanisms that solubilize sludge solids and lyse cells, thereby increasing the *rate* of degradation; 2) mechanisms that render the nondegradable organic fraction degradable, thereby increasing the *extent* of degradation; and 3) mechanisms that result in the generation of less sludge by process modification. Most sludge reduction vendors claim to do both of the first two categories. However, it is not clear that this is accomplished based on the data available from this study. Data from other investigations have indicated both an increase in the rate and the extent of degradation.

It is also not clear to what extent cell lysis accounts for increases in solids loss. For the third category, modification in the activated sludge process, usually performed on the return sludge recycle stream, can directly reduce solids (e.g., by chemical oxidation) or can solubilize solids in the same manner as Mechanisms 1 and 2, resulting in rapid degradation when the recycle stream reenters the aeration basin. A schematic for both is shown in Figures 6-1 and 6-2.

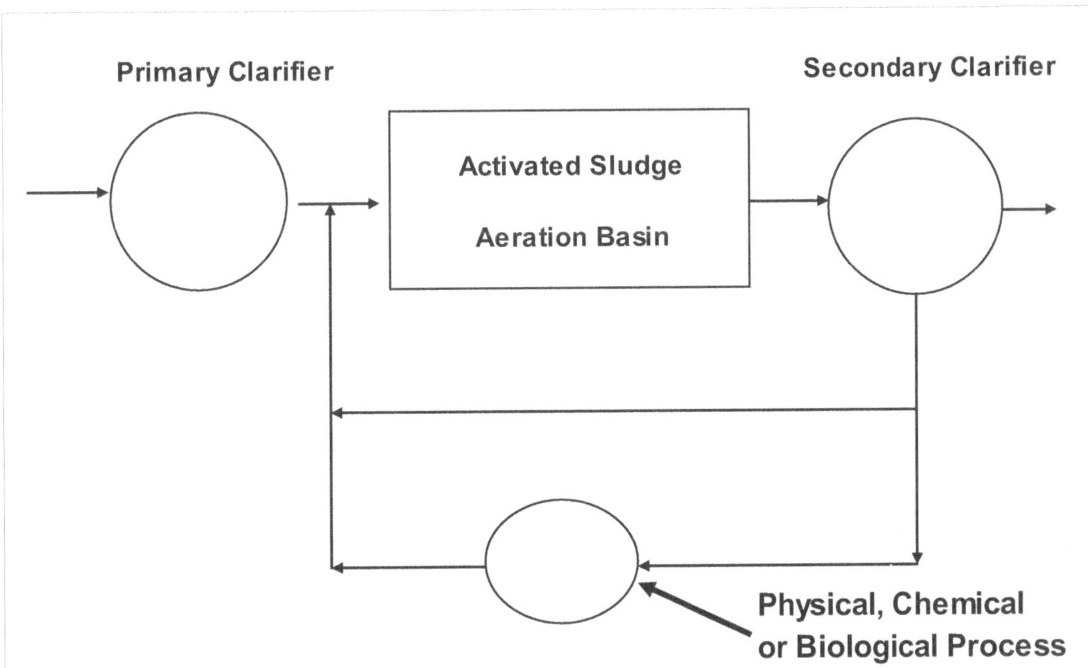

Figure 6-1. Mechanisms for Reducing Activated Sludge Solids by Process Modification.

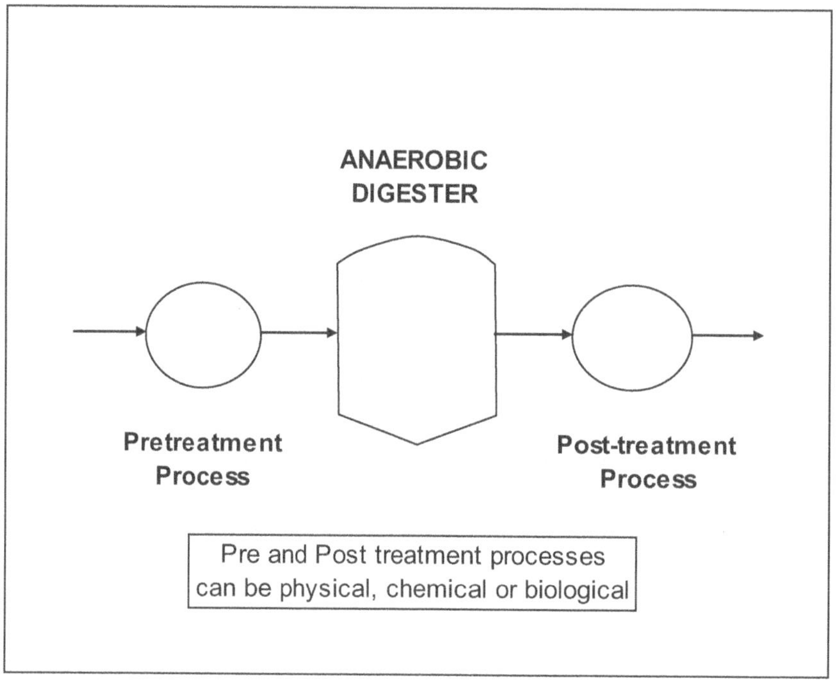

Figure 6-2. Mechanisms for Reducing Anaerobic Digester Solids by Process Modification.

This research was undertaken for several reasons. The first reason was to determine whether there were indicators either in the wastewater constituents, in the biological sludge characteristics, or in the plant operation that would provide insight into the potential for solids reduction technologies to be successful. Although it was thought that some insight into the relationship between the degree of biological solids reduction and wastewater and sludge characteristics was possible, this was considered a "long shot," primarily because the ability to obtain enough data to make comparisons across a number of plants using solids reduction technology was limited. Few solids reduction processes are in operation, and even fewer are willing to share data or samples.

The second reason for the study was to investigate mechanisms for solids reduction. This was conducted using data from both full-scale facilities as well as laboratory units where sludge reduction technologies could be compared with samples that were not treated or processed.

The third goal of the study was to develop a modeling approach that could be used to predict solids reduction if adequate data were provided. This would include distinguishing between processes that increased the rate of degradation and those that increased the extent of degradation. The model would also include processes that increased cell lysis and those that primarily solubilized particulate organic matter.

Finally, a fourth goal of this research, a goal that developed once the project was underway, was to determine the studies that would be needed to fully evaluate sludge reduction technologies in a manner that would provide results independent of vendors claims and would be universally applicable to the range of sludges encountered in practice.

6.1 Processes that Solubilize or Lyse Cells

Several processes claim to render solids more biodegradable, and most of the vendors of these processes claim that both cell lysis and solubilization of particulate solids occur, including the following:

- Mechanical shear
- Sonication
- Pressure release
- Heating under pressure
- Chemical oxidation

Most processes that solubilize or lyse cells do so by providing simple mechanical shear through intense mixing (Kady® Biolysis® Systems), sonication (Sonico), pressure release (Crown® process, MicroSludge®), heating under pressure (Cambi™ Process) or chemical oxidation, usually ozonation (Praxair®). Although various reports indicate that these processes will reduce activated sludge solids by up to 20% and increase gas production up to 30%, little independent verification has been done.

These processes can be applied either to the feed to an anaerobic or aerobic digester or in the activated sludge recycle stream. The more unusual processes are the Cannibal® process that uses an anaerobic reactor to decrease solids generation in the activated sludge process and aerobic digestion following anaerobic digestion, which provides additional solids reduction and nitrogen removal. Mechanisms for the Cannibal® process have been proposed but not confirmed. Aerobic digestion following anaerobic digestion is designed to degrade organics that do not degrade under anaerobic conditions (Kumar et al., 2006a).

Two important questions require answers with regard to implementation of these processes: How does the performance of these processes vary from plant to plant? Can we predict performance without extensive piloting? These are critical questions because field results from the Cannibal® process indicate that while most Cannibal® plants are performing satisfactorily with regard to solids reduction, a few are not performing well. Because the Cannibal® system is similar for all these plants, there must be some sludge characteristic that makes this process work better at one site than another. With the available tools and the knowledge of sludge fractionation, the research team cannot identify the characteristics that determine performance. Hence, it is recommended that facilities considering waste solids reduction technologies conduct pilot testing to vet performance. The tools provided as part of this research study can be useful as an initial step to determine the potential performance of the solids reduction technologies.

6.2 Conclusion

The state of development of sludge minimization technologies continues to advance, both for applications within the liquid treatment process and for digester pretreatment. Most technologies have full-scale applications on industrial or municipal wastewater in Europe and Australasia, with a much slower implementation rate in North America. The exceptions to this trend are the combination biological processes such as Cannibal®, which has been developed in North America and has found wider acceptance. However, many technologies have a limited number of installations (e.g., chemical treatment technologies), and some technologies have shown mixed performance, particularly on applications in North America (e.g., physical pretreatment for digestion).

This study provides valuable insight into cutting-edge research and emerging technologies associated with WAS minimization. Data analysis from several technologies indicates positive results. One of the most interesting aspects of the study is that performance of the same technology is significantly different at separate facilities. Although there is still much to understand, the results of this study improve the understanding of the mechanisms related to why some sludges are more susceptible to these technologies than others. Operators of WWTPs will find this knowledge important as they optimize systems to produce higher-quality effluent and better-quality sludge/biosolids, increase in-plant energy production, and decrease life-cycle costs.

The success of a technology depends on the understanding of the mechanisms and process parameters, particular waste stream characteristics, and economic conditions; the use of appropriate operational parameters, such as adequate input energy for physical processes; and the economics of heat, footprint, sludge disposal, existing infrastructure, equipment, chemical costs, and so forth. As the technologies evolve and more information and data are gathered, a better understanding of the efficacy and the underlying mechanisms will become clear. This study, along with other research, is a step in research that needs to continue to better understand these technologies. The information gathered through this investigation will help form the basis of that future research.

APPENDIX A

Facility Sampling and Laboratory Test Protocol

A.1 Introduction

The objective of creating a facility sampling and sludge testing protocol is to develop a set of sludge characteristics that will be used to assess the impact of sludge-reduction processes for a particular wastewater treatment facility. Sludge and wastewater samples were collected from several installations as part of this study. That experience, along with the understanding of the basic mechanistic principles behind each of the technology alternatives, served as the basis for a recommended protocol for establishing sludge characteristics for any treatment facility.

A.2 Sampling Locations and Recommended Analytical Program

Samples should be collected from numerous locations throughout the plant and analyzed for pertinent characteristics as shown in Table A-1.

Table A-1. Recommended Sampling Locations and Analytical Program.

Sample Location	pH	TS/TSS VS/VSS	COD/ sCOD	Soluble Cations/ Anions	VFAs	Total Fe/Al	Protein Polysaccharides
Raw Influent	X	TSS/VSS	X	X		X	
Primary Sludge	X	TS/VS	X	X	X	X	X
Mixed Liquor	X	TSS/VSS	X	X	X	X	X
WAS/Process Feed	X	TS/VS	X	X	X	X	X
Post Process	X	TS/VS	X	X	X	X	X
Digested Sludge	X	TS/VS	X	X	X	X	X
Dewatering Centrate/Filtrate	X	TSS/VSS	X	X		X	
Final Effluent	X	TSS/VSS	X	X		X	

Notes:
Al = aluminum
COD = chemical oxygen demand
Fe = iron
sCOD = soluble chemical oxygen demand
TS = total solids
TSS = total suspended solids
VFA = volatile fatty acid
VS = volatile solids
VSS = volatile suspended solids
WAS = waste-activated sludge

Tables A-2 and A-3 present further detail with regards to sampling location and sample volume for facilities using Cannibal™ and anaerobic digestion pre-treatment processes, respectively.

Table A-2 Activated Sludge Process Data Collection – Cannibal™.

Sample	Location	Total Sample Volume
Sample 1	Raw Influent	5.1 L
Sample 2	Primary effluent	5.1 L
Sample 3	Mixed liquor	5.1 L
Sample 4	Feed to sludge reduction process	5.1 L
Sample 5	Effluent from sludge reduction process	5.1 L
Sample 6	Final Plant Effluent	5.1 L

NOTE:
L = liters

Table A-3. Digester Pre-treatment process Data Collection.

Sample	Location	Sample Volume
Sample 1	Raw Influent	5.1 L
Sample 2	Primary sludge	5.1 L
Sample 3	Waste activated sludge	5.1 L
Sample 4	Feed to sludge reduction process (if different from Samples 2 and 3)	5.1 L
Sample 5	Sludge discharge from the sludge reduction process	5.1 L
Sample 6	Digested sludge	5.1 L
Sample 7	Dewatered cake	5.1 L
Sample 8	Dewatering filtrate/centrate	5.1 L

NOTE:
L = liters

Analyses should be conducted according to the methods described in Table A-4. All samples for soluble fraction analyses (sCOD, cations, anions, proteins, polysaccharides, and VFAs) should be centrifuged and the centrate should be filtered through 0.45-micron nitrocellulose membrane filters.

Table A-4 Recommended Sample Analyses and Corresponding Methods.

Analysis	Method	Reference
pH	Accumet electrode with AgCl filling solution	N/A
TS/VS and TSS/VSS	Standard Methods 2450 A, B, D, and E	(A.P.H.A. et al., 1995)
Total and Soluble COD	Hach High Range COD tubes (20–1,500 mg/L COD)	(Hach Company, Loveland, CO)
Soluble Cations	Dionex DX-120 Ion Chromatograph with IonPac either CS16 5x250 mm or CS12 5x250 mm columns, CG16 5x50 mm guard column and either 25 mM methane sulfonic acid or 10 mM sulfuric acid as the eluents	N/A
Soluble Anions	Dionex DX-120 Ion Chromatagoraph with IonPac AS9-HC 4x250 mm column, AG9-HC 4x50 mm guard column and 9 mM sodium bicarbonate as the eluent	N/A
VFAs	HP 5890 with flame ionization detector and Nukol Silca Capillary Column, 15 m x 0.53 mm x 0.5 mm film thickness; carrier gas pressurers: He= 17 mL/min; N2= 13 mL/min; Air= 450 mL/min; H2= 45 mL/min; column pressure= 15 psi	N/A
Total Aluminum	Perkins-Elmer 5100 Flame Atomic Adsorption Spectrophotometer; gases= nitrous oxide; acetylene; wavelength= 309.3 nm; method detection limits= 5–100 mg/L	N/A
Total Iron	Perkins-Elmer 5100 Flame Atomic Adsorption Spectrophotometer; gases= air, acetylene; wavelength= 248.3nm; method detection limits= 0.3–10 mg/L	N/A
Protein	Frolund et al. modification of Lowry method, using BSA and humic acid as standards; Bradford protein assay using BSA as standard	(Bradford, 1976; Frolund et al., 1995)
Polysaccharides	Colorimetric analysis using sulfuric acid and phenol with glucose as standard	(Dubois et al., 1956)

Notes:
BSA = bovine serum albumin
COD = chemical oxygen demand
H_2 = hydrogen
He = helium
L = liter
m = meter
mg = milligram(s)
min = minute
mL = milliliter(s)
mm = millimeter(s)
mM = micromoles
N_2 = nitrogen gas
nm = nanometer
N/A = not applicable
psi = pounds per square inch
TS = total solids
TSS = total suspended solidsa
VFA = volatile fatty acid
VS = volatile solids
VSS = volatile suspended solids

A.2.1 Sample Processing Methodology

Depending on the analysis being performed on a sample, certain preservation procedures should be implemented. All samples should be shipped on ice to hinder additional biological activity. For iron and aluminum analysis, a volume of sample should be preserved using concentrated nitric acid. For COD analysis, a volume of sample should be preserved with concentrated sulfuric acid. Upon arrival to the laboratory, a volume of the non-acidified samples should be centrifuged at approximately 17,700 x G on a Beckman-Coulter Avanti-JE high-speed or similar refrigerated centrifuge for 15 minutes; the centrate filtered through a 0.45-micron nitrocellulose membrane filter and frozen until analysis for soluble COD, protein, polysaccharides, cations, anions, and VFAs can be performed. All other samples should be stored in a 4°C refrigerator until analysis. Analysis of samples should be conducted within a week of arrival, with priority given to those procedures that required shorter hold times as specified by Standard Methods (APHA, 1998).

APPENDIX B

EVALUATION OF THE MICROSLUDGE™ DEMONSTRATION PROJECT AT THE DES MOINES, IOWA WATER RECLAMATION FACILITY

B.1 Introduction

This report is a supplemental component of the Water Environment Research Foundation's *Evaluation of Processes to Reduce Activated Solids Generation and Disposal* (05-CTS-3) project that aims to establish a comprehensive evaluation methodology for the various waste activated sludge (WAS)-reduction technologies available. A case study in the application of the MicroSludge™ sludge pretreatment technology at the Des Moines, Iowa, Metropolitan Wastewater Reclamation Authority's Wastewater Reclamation Facility (WRF) is presented and discussed here. The Des Moines WRF services approximately 400,000 population equivalent and generates about 5,000 dry tonnes of WAS annually. A MicroSludge™ process (provided by Paradigm Environmental Technologies, Inc.) was installed as part of a full-scale demonstration study in late 2007 to evaluate its ability to increase biogas production and decrease the quantity of digested residuals for disposal. At the WRF, MicroSludge™ processing of thickened WAS (TWAS) followed by mesophilic co-digestion with primary sludge (PS) and trucked liquid waste (TLW) was tested at full scale. Co-digestion of MicroSludge™ conditioned TWAS with PS and digestion of conditioned TWAS only was tested at the pilot scale.

The specific scope of services for this additional component of the project is as follows:

Task 1. Review and Evaluation of Existing Data. This was conducted by the 05-CTS-3 project team acting as an independent third party, using data collected by Paradigm Environmental Technologies previously for the full-scale studies and the new data collected as part of the additional pilot-scale trials.

Task 2. Expert Advice during Continued Testing. Partial financial support by WERF allowed the City of Des Moines and Paradigm Environmental Technologies to continue their testing through August 2009. WERF funding was used to cover the labor and expenses associated with advisory services provided by the 05-CTS-3 project team members during this additional testing period. This task included the review of the proposed monitoring program of the pilot-scale anaerobic digestion and residual's odor potential and dewatering study components.

Task 3. Partial Funding of Additional Analytical Program. WERF provided partial funding required for the analytical programs conducted by independent laboratories and other research institutions for the continuation and expansion of the MicroSludge™ demonstration study at Des Moines.

Task 4. Preparation of Independent Report. The 05-CTS-3 project team members prepared an independent report summarizing the observations and findings of the Des Moines MicroSludge™ demonstration study, covering both trial stages.

This report corresponds to Task 4, and provides a summary of the evaluation conducted on the data received from Paradigm Environmental Technologies on their demonstration using MicroSludge™ at the Des Moines WRF, Iowa. The demonstration study included both the full-scale and the pilot-scale components, which were conducted between November 2007 and August of 2009.

B.2 Description of the MicroSludge™ Process

MicroSludge™ (developed by Paradigm Environmental Technologies) is a chemical and mechanical pre-treatment process that conditions TWAS to increase both the rate and the extent that TWAS is degraded in an anaerobic digester. The process uses caustic pre-treatment to weaken cell membranes, followed by a high-pressure homogenizer, or "cell disrupter," to provide a large and sudden pressure drop to burst the cells in TWAS. The premise of the process is that by rupturing cell membranes and releasing the cytoplasm of TWAS, the concentration of soluble solids from TWAS is greatly increased, potentially enabling anaerobic digesters to convert a larger portion of TWAS to biogas and at a faster rate (Stephenson and Dhaliwal, 2000). A process flow diagram of a typical installation of a MicroSludge™ *cell* disrupter and a central processing skid with the unit processes and pumps is illustrated in Figure B-1.

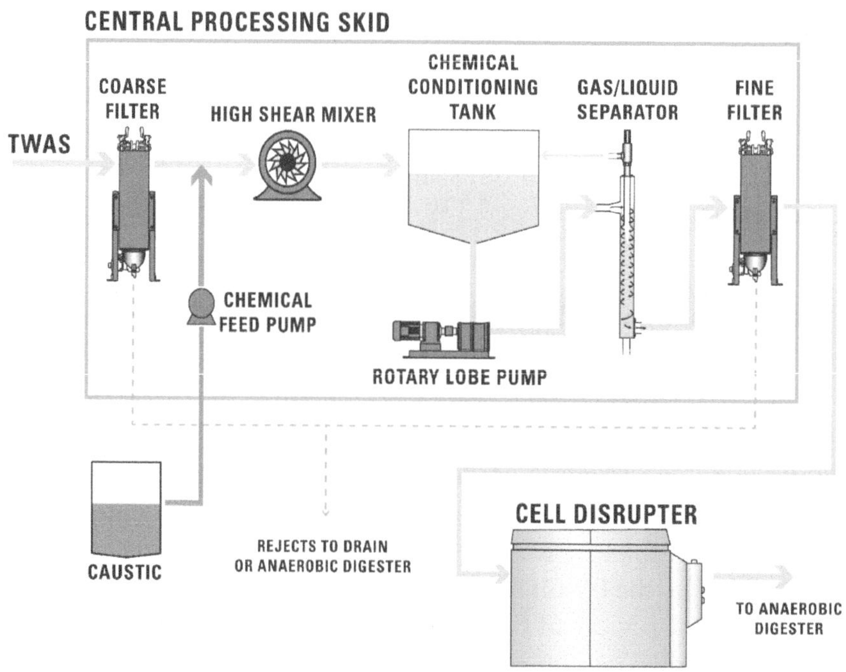

Figure B-1. MicroSludge™ Process Flow Diagram.
Reprinted with permission from Stephenson et al., 2009.

The benefit of increased rate of WAS degradation is the ability to achieve the same digester performance in lesser time, and therefore minimize reactor volume, while the benefit of increasing the extent of WAS degradation is the increase in conversion of biomass to biogas, thus reducing the mass of biosolids for further processing and ultimate disposal, and increasing hydrocarbon availability for renewable energy production. Other potential secondary benefits include decreased anaerobic digester foaming, decreased biosolids odors, and better cake dewatering capability. It should be noted that these benefits have been observed in similar anaerobic digestion WAS conditioning processes, such as pressure release, sonication, pulse electrical field, and thermal hydrolysis (Sandino et al., 2008).

B.3 Synopsis of Previous Studies

The first full-scale evaluation of the MicroSludge™ process was conducted at the Chilliwack Wastewater Treatment Plant (WWTP) in British Columbia. Results of this study indicated that this technology increased the rate and extent of TWAS degradation in a conventional mesophilic anaerobic digester (Rabinowitz and Stephenson, 2005). When compared with a control digester fed with 65% primary sludge and 35% untreated TWAS (by mass), operating at a hydraulic retention time (HRT) of 13 days; the test digester fed with MicroSludge™ pre-treated TWAS operating under identical conditions exhibited 18% increase in average volatile solids reduction (VSr). It was also concluded that the retention time required for anaerobic digestion could be reduced from 18 days to between 12 and 15 days.

Following the demonstration of the MicroSludge™ unit at the Chilliwack WWTP, a second full-scale unit was installed at the Los Angeles County Sanitation Districts' Joint Water Pollution Control Plant (JWPCP). The MicroSludge™ pre-treated TWAS was mixed with primary sludge with an approximate 32:68 mass ratio and co-digested at an HRT of 19 days. According to Rabinowitz and Stephenson (2006), the increase in VSr when compared to the control digester was not significant. The VSr in the control digester ranged from 51 to 54%, while the test digester receiving MicroSludge™-processed WAS ranged from 55 to 58%. Biogas generation data also exhibited a similar trend as VSr data, although considerable variability in gas volumes was observed over time. However, laboratory studies conducted at JWPCP using 100% MicroSludge™-treated TWAS exhibited 8% increase in VSr when compared with a control digester. One advantage exhibited by the laboratory setup was that it reduced any foaming, which led to the conclusion that the MicroSludge™ technology was capable of destroying foam forming filamentous microorganisms.

B.4 Full-Scale Demonstration Study Component

B.4.1 General Description and Experimental Setup of Study Component

Two MicroSludge™ cell disrupters (with one unit serving as a "hot" standby) and a central processing skid with controls, instrumentation, tanks, screens, and pumps were installed at the WRF. Each cell disrupter uses a 150 horsepower (hp) motor, a three-plunger pump, a cell disruption valve, and associated controls. Each cell disrupter can process about 67 liters per minute (18 gallons per minute [gpm]) of TWAS at a maximum pressure of 99,970 kilopascal (kPa) (14,500 pounds per square inch [psi]). Figure B-2 presents a photograph of the units installed at Des Moines.

Figure B-2. MicroSludge™ Installation at the Des Moines WRF.
Reprinted with permission from Stephenson et al., 2009.

Starting in November 2007, MicroSludge™ was operated at the WRF to process 96 cubic meters per day (m^3/day) (nominally 26,000 gallons per day [gal/day]) of rotary drum-thickened WAS at a total solids (TS) concentration between 6 and 10%. In two "test" digesters, the MicroSludge™-processed TWAS was co-digested in 2.25-million-gallon mesophilic anaerobic digesters with PS and TLW in a TWAS:PS:TLW ratio of approximately 14:33:53 by mass loading of chemical oxygen demand (COD) at an HRT of 36 days. In two identical "control" digesters, raw TWAS was co-digested in a TWAS:PS:TLW ratio of approximately 25:29:46 by mass loading of COD at an HRT of 26 days. Figure B-3 illustrates the average digester feed distribution to WRF anaerobic digesters (Stephenson et al., 2009).

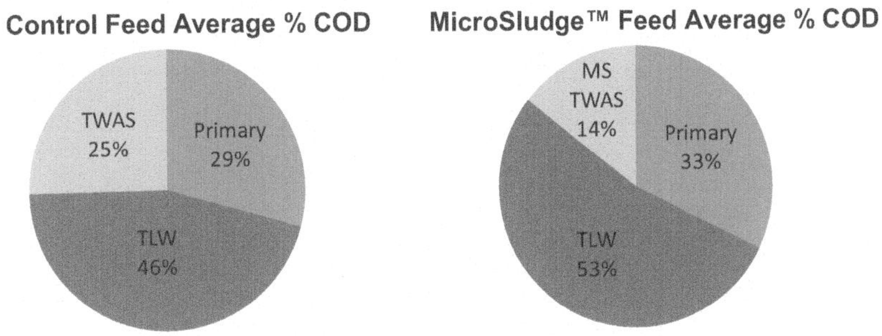

Figure B-3. Average Feed Distribution during Full-Scale Testing of MicroSludge™ at Des Moines WRF. Reprinted with permission from Stephenson et al., 2009.

The relative small amount of MicroSludge™- processed TWAS fed to the test digester, as well as the highly variable daily contribution of TLW (ranging from 20-80% of the total COD daily influent load), and the unequal feed characteristics between the test and the control digesters made the full-scale digester performance data difficult to identify statistically significant performance improvements. However, the demonstration study did provide valuable information regarding the operational requirements, and the reliability of the full-scale electromechanical and instrumentation components of the cell disruptor and the central processing skid system provided by Paradigm. This report concentrates on these particular aspects of the demonstration study.

B.4.2 MicroSludge™ Full-Scale Maintenance and Reliability Assessment

The Des Moines WRF full-scale demonstration study provided an opportunity to test various MicroSludge™ component designs and to maximize component life under relatively harsh operating conditions. During 2008, extensive flooding in the area resulted in a significant increase in abrasive inorganic material in the sludge for 4 months (Figure B-4). While being an unfortunate circumstance for measuring digester performance, this provided a suitable environment for carrying out accelerated life testing on the MicroSludge™ equipment. Wear components were operated to failure, instead of operating the equipment with regular maintenance to avoid downtime. This allowed side-by-side comparison of different component designs.

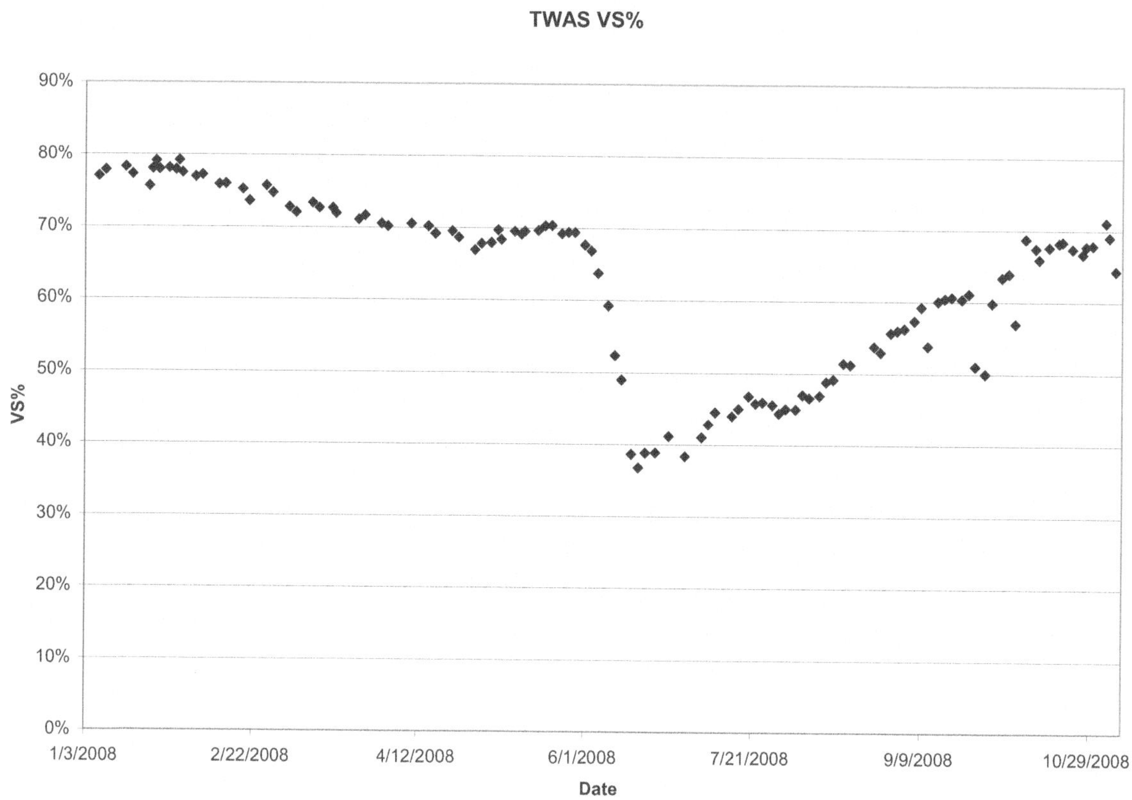

Figure B-4. Volatile Solids Variability in TWAS-Fed to MicroSludge™ System at the Des Moines WRF.

System availability information is presented in Figure B-5. The chart shows MicroSludge™ system availability from January to September 2008 from before and during the accelerated life testing compared with the uptime in 2009 when all the design improvements had been implemented.

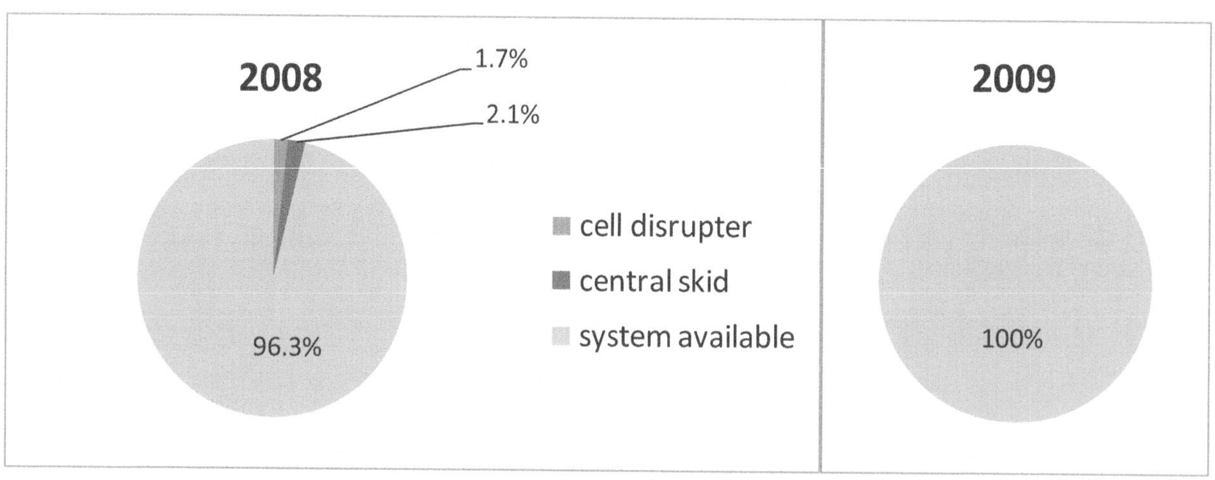

Figure B-5. MicroSludge™ System Reliability at the Des Moines WRF.

As noted in this report, the MicroSludge equipment was operated to failure to determine the required maintenance frequency of various parts to optimize their design under the abrasive conditions of WAS with an extraordinarily high-silt content during the flood in Des Moines. Periodic interruptions in sludge processing were a consequence of this investigation.

Components that caused downtime were those listed in Table B-1. Replacement of cylinder packing, check valve balls, and cell disruption valves accounted for the downtime of the cell disruptor. These components are normal wear items that require routine replacement according to scheduled maintenance.

In addition, a failure of two analog input cards occurred for the programmable logic controller (PLC). The PLC card failure was due to corrosion of the electronics. The PLC electronics were not conformably coated and therefore at risk of corrosion due to hydrogen sulfide gas. Future systems shall use conformably coated electronics, eliminating this mode of failure.

Typical downtime for any one of the failures was similar to the time listed as labor hours for maintenance plus often additional time for travel to site if failures occurred when no one was on site. Downtime was not tracked per component.

Based on the information collected from the Des Moines WRF demonstration study, as well as from experienced gathered elsewhere, the manufacturer of MicroSludge™ has developed an estimate of the expected maintenance requirements for a full-scale installation of similar processing capacity as the one tested, which is presented in Table B-1.

Table B-1. Maintenance Schedule for a Full-Scale MicroSludge™ System.

Component	Category	Period (weeks)	Labor (hours)	Total Annual Labor (hours)
Cylinders	Packing	4	2	26
Cylinders	Gland	52	1	1
Check Valves	Balls	8	2	13
Check Valves	Valve Seats	52	2	2
Homogenizing Valves	Replacement	4	1	13
Homogenizing Valves	Block	52	2	2
Filters	Cleaning	52	8	8
Pumps	Lobe and Wear plate	26	2	4
Cell Disrupter	Lubrication	26	4	8
General Maintenance	Cleaning and Lubrication	52	8	8

The high level of inorganic material in the WAS at the Des Moines WWTP also provided an opportunity to test the effectiveness of a de-silting hydrocyclone. Figure B-6 shows a schematic of a hydrocyclone that was installed on the WAS inlet pipe to the rotary drum thickener. The hydrocyclone was used to remove and concentrate inorganic material and direct this flow to the RDT filtrate. Table B-2 shows the total and volatile solids in the feed and reject stream from the hydrocyclone.

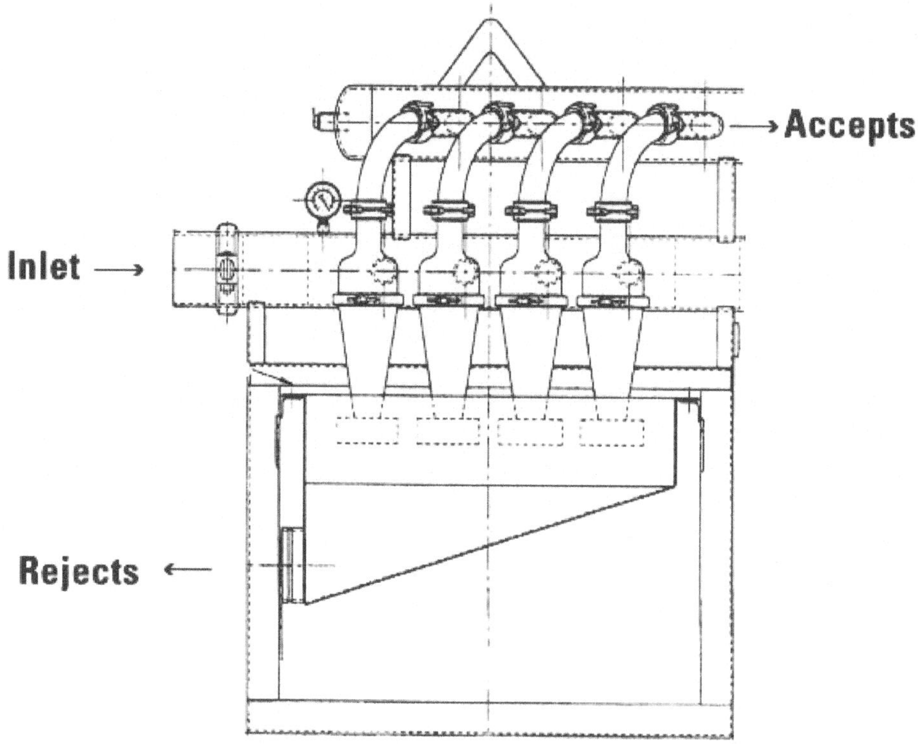

Figure B-6. Desilter Hydrocyclone Configuration, Tested at the Des Moines WRF.

Table B-2. Hydrocyclone Performance at the Des Moines WRF.

Hydrocyclone Operations		Inlet		Reject	
	Units	Average	St. Dev.	Average	St. Dev.
Approximate Flow	gpm	400		0.5	
Total Solids	mg/L	15,554	476	42,606	12,736
Volatile Solids	mg/L	10,854	701	21,703	3,952
Fixed Solids	mg/L	4,700	304	20,903	8,986
% Volatile Solids	%	70	2	52	6
% Fixed Solids	%	30	2	48	6

The hydrocyclone was effective in removing a portion of the inorganic solids in the WAS. The cyclone was able to concentrate the inorganic portion of the sludge by approximately 4 times. The volatile solids concentration also increased by two times the inlet concentration. This shows that the cyclone was effective in providing a distinct separation of the inorganic material in the WAS, as opposed to simply concentrating the entire influent solids.

The hydrocyclone required a minimum inlet pressure of 35 psi to maintain effective separation. Since it was not possible to consistently provide the required inlet pressure while maintaining the desired wasting rate from the secondary treatment plant at the Wastewater Reclamation Authority, the hydrocyclone system was bypassed once the concept of grit removal had been proven.

Installation of the hydrocyclone closer to the WAS pump outlet would have allowed for more consistent operation, but due to space constraints and other physical limitations, it was not practical to locate the hydrocyclone near the secondary treatment plant. In a permanent hydrocyclone installation, a dedicated reject collection and drainage system would be recommended to remove the grit from the digester feed.

B.4.3 Conclusions and Recommendations

The full-scale demonstration study at the Des Moines WRF, although inconclusive in terms of identifying significant performance improvements primarily due to issues related to feed characteristics, did offer an opportunity to assess the overall reliability and maintenance requirements of a full-scale MicroSludge™ installation.

Results from the demonstration study indicate that to maximize system reliability, a design of a MicroSludge™ system should include the following measures:

- Running a scheduled maintenance program to replace wear components before failure.
- Incorporating redundant screens, macerator and cell disruptor (CD) feed pump, all on hot standby.
- Minimizing CD feed pressure to reduce pump rotation speed and correspondingly decrease wear.
- Optimized design of CD valve and valve block.
- Optionally including hydrocyclones to remove grit from WAS.

B.5 Pilot-Scale Demonstration Study Component

B.5.1 General Description of Study Component

Due to the variability of PS, TWAS, and TLW fed to the digesters, and the differences between the control and the test digesters during the full-scale component of the demonstration study, pilot-scale anaerobic digesters were designed and installed to generate additional digester performance data under more stable loading conditions. The objective of the pilot-scale study was to understand the effect on the performance of anaerobic digesters using MicroSludge™-processed sludge and its effect on the quantity and quality of biogas produced. The main differences between this and the previous full-scale trial study component were the elimination of TLW in the feed, the ability to maintain consistent feed quality of the other sludges, and the evaluation of the effects of reducing the HRT for anaerobic digestion of MicroSludge™-processed sludge.

Four 312-liter continuously mixed mesophilic anaerobic digesters were setup at the Des Moines WRF. One setup was targeted to evaluate the anaerobic digester performance using a control and a test digester fed with a mix of PS and MicroSludge™-processed TWAS, operating at varying HRTs, while the other setup was targeted to evaluate performance using a control and a test digester fed with TWAS only, operating at a 15-day HRT. A summary of the experiments performed using the pilot scale mesophilic anaerobic digesters is presented in Table B-3. Figure B-7 presents a photo of the pilot digesters (Stephenson et al., 2009).

Table B-3. Pilot Scale Anaerobic Digester Experiments.

Digester Number	Digester Feed (TS Basis)	HRT (Days)	Organic Loading Rate (Kilograms Volatile Solids/m³ day)
1	56% TWAS + 44% PS	20, 15, 10, 7	1.45, 1.76, 2.68, 4.10
2	56% MicroSludge™ TWAS + 44% PS	20, 15, 10, 7	1.45, 1.76, 2.68, 4.10
3	100% raw TWAS	15	1.83
4	100% MicroSludge™ TWAS	15	1.83

The following sections summarize the pilot-scale digester experimental setups, the analyses performed, and the corresponding results and conclusions that can be derived from these trials.

Figure B-7. Pilot-Scale Digesters at Des Moines WRF.

B.5.2 Co-Digested PS and TWAS Operating at Varying HRTs

Two pilot-scale digesters were operated as part of these trials; one as the control and the other as the test digester. The control digester was fed with a mixture of PS and untreated TWAS, while the test digester was fed with a mixture of PS and MicroSludge™-processed TWAS. A schematic of the pilot-scale setup is presented in Figure B-8. In both of the digesters, the PS made up 44% and the TWAS made up 56% of the TS in the feed mixture. Both digesters were operated concurrently, starting with an HRT of 20 days and reducing to 15 days, 10 days, and 7 days. The number of days of operation, HRT, and the dates are included in Table B-4.

Table B-4. Duration of Testing Conditions of the Pilot-scale Co-Digestion Study.

HRT	Control Digester (2009)	Test Digester (2009)	No. of days
20 days	December 20 – March 15	December 20 – March 15	86
15 days	March 16 – April 1	March 16 – April 1	17
10 days	April 2 – May 22	April 2 – May 22	51
7 days	May 23 – August 7	May 23 – August 7	77

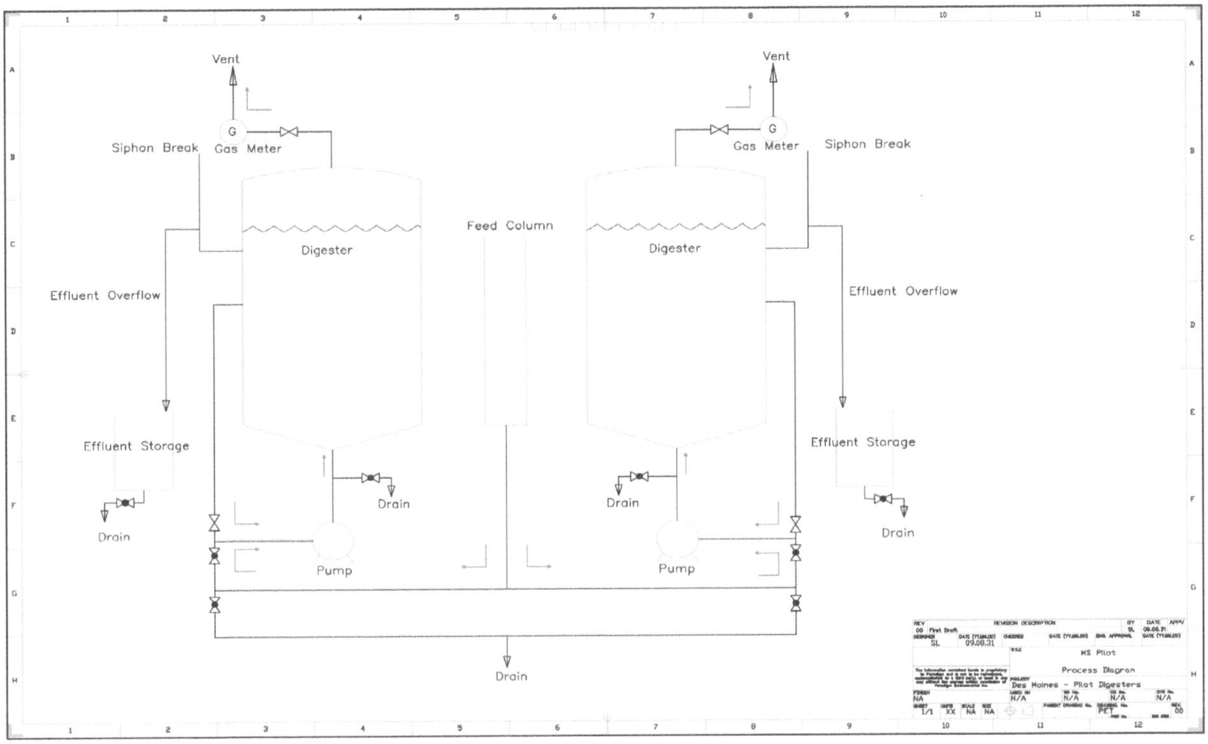

Figure B-8. Pilot-Scale Setup of the Control and Test Anaerobic Digesters Fed with a Mix of Primary and Secondary Sludge at Des Moines WRF.

During the operation, total and volatile solids were measured as per Standard Methods (APHA, 1998) in addition to COD, pH, and alkalinity for the control and test feed and effluent. Digester effluent volume, digester temperature, and volume of biogas produced in each digester were also measured on a daily basis.

A pair of Ritter brand wet test gas meters were used to measure the gas flow of the control and the test digesters. The gas meters were calibrated before use. Carbon dioxide was analyzed using a Bacharach flue gas analyzer. Gas samples were also sent out to be analyzed for methand (CH_4), oxygen (O_2), and nitrogen (N_2) by Marquette University.

Each morning, the digester temperature, effluent volume, and the gas produced overnight was recorded. Samples of PS, raw TWAS, and MicroSludge™-processed TWAS were collected for TS analysis using a microwave oven moisture analyzer. These TS values were then used to dilute each type of sludge to a consistent TS level (PS was diluted to 4% TS and TWAS were diluted to 5% TS). Dilution water for the PS was the primary effluent from the primary clarifiers and dilution water for the TWAS was filtrate from the rotary drum thickeners.

After the sludge was diluted to appropriate levels, each type of sludge was fed separately into the appropriate digester. Only during the first and second phases of the experiment (20 and 15 days HRT), the primary and secondary sludge were mixed together and fed at the same time. Outside of the 20- and 15-day HRT regimes, the sludge mix was fed separately, typically with the primary sludge being fed into the digesters first. This was due to the size of the feed column (about 22 liters) and the volume of sludge required (more than the volume of the feed column).

The digesters were mixed semi-continuously using an air diaphragm pump. The recycle loop allowed for the digester contents to be removed from mid-height of the digester and to be reintroduced into the bottom of the digester. Heating of the digesters was maintained by two 300-watt belt heaters on each digester, controlled via an Omega thermostat. The digester temperature averaged 37°C and was monitored by a thermocouple.

Samples of raw and anaerobically digested sludge from control and test digesters were sent to Dr. John Novak at Virginia Tech for organic sulfur odor testing and batch digestion studies to determine the rates of change in volatile fatty acid (VFA) generation, soluble phosphorus production, and organic nitrogen speciation via five-day incubation. Samples of untreated and MicroSludge™-treated sludge samples were also sent Dr. Matthew Higgins at Bucknell University for total coliform, fecal coliform, and *E. Coli* analysis using United States Environmental Protection Agency (U.S. EPA) Method 1680.

B.5.3 Results and Discussion

B.5.3.1 Effect of Activated Sludge Operating Conditions on Pilot Study

Table B-5 summarizes the operational conditions of the WRF activated sludge process during the various phases of the experiment. The activated sludge solids retention time (SRT) and operating temperature will ultimately determine the biodegradability of the WAS, and thus the extent of its VSr and COD reduction (CODr) during anaerobic digestion. Because the activated sludge SRT and operating temperature varied temporally as the pilot study was being conducted, care should be taken when comparing results and drawing conclusions between different anaerobic digestion HRTs.

Table B-5. Activated Sludge Operational Conditions during MicroSludge™ Study.

Date	MLSS (mg/L)	MLVSS (%)	Aeration Basin Temperature (oC)	SRT (days)	Anaerobic Digestion HRT (days)
December 2008	4,534	75	15	23	20
January 2009	4,251	77	14	21	
February 2009	4,386	79	14	22	
March 2009	4,897	75	14	18	15
April 2009	5,009	77	15	15	10
May 2009	4,854	73	17	16	
June 2009	4,439	77	19	14	7
July 2009	3,659	78	20	14	
August 2009	4,056	78	21	12	

B.5.3.2 Organic Sulfur Odor Testing

Odor potential testing was conducted at Virginia Tech on samples provided from the MicroSludge™ study at the Des Moines WRF. Figure B-9 shows the total volatile organic sulfur (TVOSC) measured from sludge samples obtained from control and test digesters during 5 days incubation.

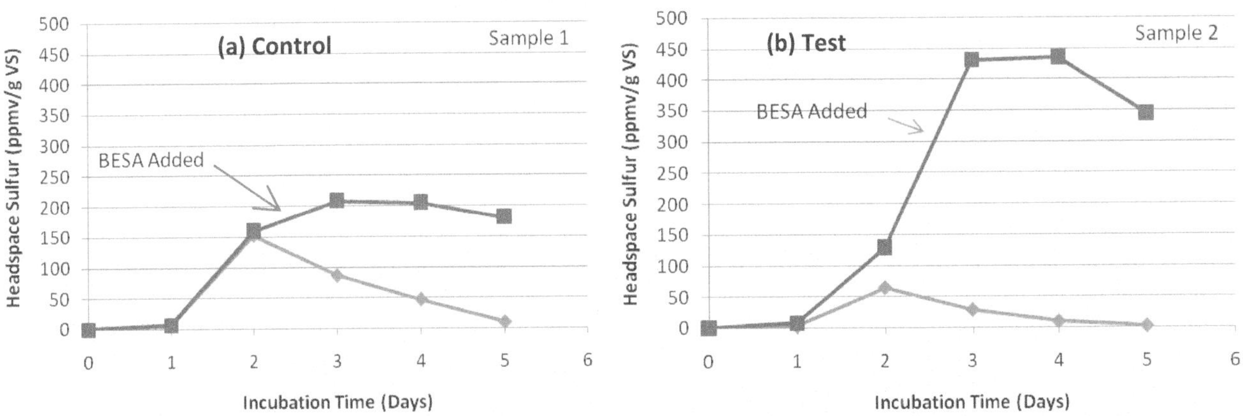

Figure B-9 (a) TVOSC for Control Sample Before and after Bromoethane Sulfuric Acid (BESA) Addition, (b) TVOSC for Test Sample Before and After BESA Addition.

The data presented in Figure B-10 indicates that the organic sulfur emission from MicroSludge™-treated sludge was about one-third (50 versus 150 parts per million by volume per gallon [ppmv/g] volatile solids) of that of the control digester on the second day of incubation. This points toward less odor generation from MicroSludge™-treated sludge. Also as part of the trials, BESA was added to samples of the digesters to test the methanogenic activity in the digesters. On the third day of incubation, the TVOSC for the BESA-amended sludge was 210 ppmv/g volatile solids for the control digester, and 440 ppmv/g volatile solids for the test digester. Addition of BESA-to-sludge samples results in the inhibition of methanogens, and thus the BESA-amended data indicates that the methanogenic population obtained from the test digesters was more active than for the control.

B.5.3.3 Batch Digestion Studies

Untreated and MicroSludge™-pretreated sludge samples provided to Virginia Tech were incubated for 5 days to determine the rates of change in VFA generation, soluble phosphorus production, and organic nitrogen speciation. These parameters would indicate the extent of solubilization of the biomass as a result of the MicroSludge™ process. It is unclear if BESA was added to these samples. If so, then the VFA levels are VFA potential. If not, then VFA levels are net VFA generation less VFA consumption by the MPB.

The data obtained for VFA generation are shown in Figures B-10a and B-10b, which include total VFA, acetic acid (HAc), proprionic acid (HPr), and butyric acid (HBu). Figure B-10a indicates the VFAs in the untreated sludge sample were rapidly generated over a day of incubation, which then slowly increased to a maximum of 17,600 mg/L over the next three days. In comparison, the MicroSludge™-processed sludge sample (Figure B-10b) contained 15,000 mg/L VFA at the laboratory following overnight shipping on ice. The VFA concentration continued to increase to almost 20,000 mg/L over the next five days during incubation. This indicates that MicroSludge™ pretreatment can lead to the direct formation of VFAs due to solubilization of sludge samples. The presence of significantly higher amounts of VFAs will consequently reduce the reaction time required for anaerobic digestion, since hydrolysis and acidogenisis (the necessary preceding steps before methanogenesis) have already occurred. By comparison, an acid-phase anaerobic digestion system is typically designed for a 2-day SRT; hence, this data suggests that the rate of digestion is significantly increased by the MicroSludge™ process.

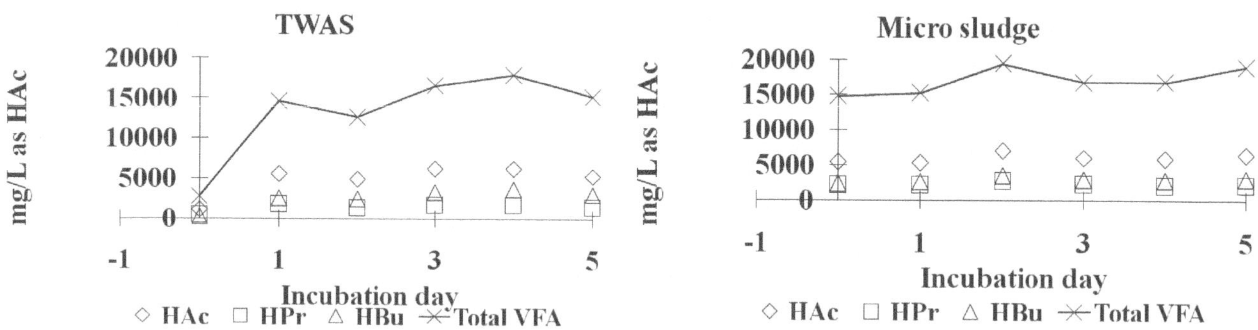

Figure B-10 (a) Production of VFAs from Untreated TWAS, (b) Production of VFAs from MicroSludge™-Processed TWAS.

The soluble phosphorus released from untreated and MicroSludge™-pretreated sludge samples were monitored and plotted as shown in B-11. The soluble phosphorus following MicroSludge™ processing was observed to be lower than the soluble phosphorus for the untreated sludge sample. This could be due to chemical precipitation of phosphorus. Phosphorus can precipitate as calcium phosphate at higher pH values, which could be the result of the caustic addition as part of the treatment process.

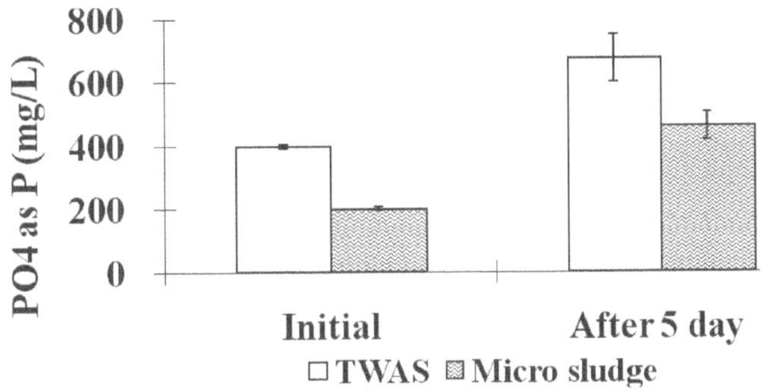

Figure B-11. Soluble Orthophosphorus in Control and Test Samples Before and After 5 Days Incubation.

Figures B-12a and B-12b show the various species nitrogen compounds in control and test sludge samples over 5 days of incubation. The speciation of nitrogen compounds was observed to be very different in the initial control and test sludge samples. The initial total Kjeldahl nitrogen (TKN) for the untreated TWAS was primarily organic nitrogen, with ammonia nitrogen constituting only about 8% of the TKN. For the MicroSludge™-processed TWAS, however, the TKN was somewhat less when compared to the control sample (indicating some nitrogen loss with MicroSludge™ processing probably due to ammonia volatilization at the higher pH induced by the caustic addition), and the ammonia fraction was much higher, constituting about 38% of the TKN, showing a rapid conversion of organic nitrogen to ammonia. After five days incubation, the overall quantities and relative distribution between ammonia and organic nitrogen in both the sludge samples were similar.

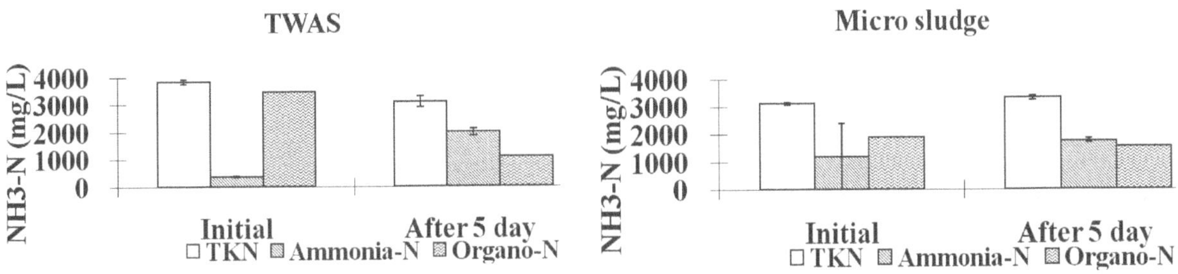

Figure B-12 (a) Nitrogen Distribution for Untreated TWAS Before and After 5 Days Incubation, (b) Nitrogen Distribution for MicroSludge™-Processed TWAS Before and After 5 Days Incubation.

Results from these anaerobic incubation tests would indicate that the MicroSludge™ treatment of TWAS did not increase the levels of soluble phosphorus in the digester content, probably as the result of the higher pH levels caused by the caustic addition step that is part of the conditioning process. Increase in the nutrient content of digested sludge reject streams (such as those generated from subsequent dewatering steps) are one of the negative impacts normally associated with the pre-treatment of TWAS ahead of anaerobic digestion (Sandino et al., 2008).

B.5.3.4 VS, COD, and TPS Reduction

Per EPA *Code of Federal Regulations* (CFR) Part 503 Regulations, the objectives of anaerobic digestion are to stabilize wastewater residuals for safe reuse. The intent of using anaerobic digestion in lieu of other residuals stabilization processes are the significant economic, social, and environmental benefits, which include biosolids mass reduction and the production of a methane-rich biogas yielding the potential for renewable energy production.

To provide a more comprehensive analysis, this research considered both VSr and CODr to assess digester performance. Table B-6 summarizes the VS data obtained from the control and the test digesters fed with a mix of PS and TWAS. Figure B-13 shows the trend in VSr at different HRTs for the control and test digesters. From the data, maximum VSr in both digesters was observed at 20-day HRT; however, at this HRT, VSr efficiency for both digesters was essentially the same. Overall VSr was also observed to be higher at the 7-day HRT when compared to 15 and 10-day HRTs. This may be due the shorter activated sludge process SRT at which the WRF was being operated (at 12 days, the shortest of all during testing as indicated in Table B-6), which would result in a higher biodegradable fraction in the VSr when compared to the control digester. The VSr results would appear to indicate that this technology may not be of much advantage to anaerobic digesters operating at or above 20 days HRT, and that its potential to increase the extent of biomass reduction is only achievable at lower HRTs. This result is consistent with the previous study conducted at Chilliwack WWTP and suggests that MicroSludge™ is capable of increasing the rate of digestion and, therefore, may reduce the retention time (thus digester volume) required for anaerobic digestion.

Table B-7 summarizes the COD data obtained from the control and the test digesters. Figure B-14 shows the trend in CODr at different HRTs for the control and test digesters. Similar to volatile solids, the COD concentration in the feed to both the digesters was maintained same. A very similar trend was observed between CODr and VSr, showing potential benefits of the MicroSludge™ process in terms of organic matter reduction only when digester HRTs are kept below 20 days.

Table B-8 summarizes the total particulate solids (TPS) data obtained from the control and the test digesters, and, Figure B-15 shows the trend in TPS reduction (TPSr) at different HRTs for the control and test digesters. At 20 days HRT, maximum TPSr was observed in both the digesters. However, at this HRT, unlike VS and COD data, there was a 14% increase in TPSr in the test digester when compared to the control digester. This TPSr is a result of solubilization of residuals by the MicroSludge™ process using physical and chemical mechanisms. The trend was very similar to the trend observed between CODr and VSr, showing a maximum increase in TPSr in the test digester at 10-day HRT.

Table B-6. Volatile Solids Data Obtained from the Control and the Test Digesters.

HRT	No. of Data points	Control Feed Volatile Solids (mg/L)	Test Feed Volatile Solids (mg/L)	Control Effluent Volatile Solids (mg/L)	Test Effluent Volatile Solids (mg/L)	% VSr in Control Digester	% VSr in Test Digester	% Increase in VSr in Test
20-day	30	31,503	31,503	15,886	15,748	50	50	0
15-day	7	27,119	27,119	17,480	16,864	36	38	5
10-day	19	27,717	27,717	18,067	16,296	35	41	15
7-day	24	28,585	28,585	17,532	15,707	39	45	13

Table B-7. COD Data Obtained from the Control and the Test Digesters.

HRT	No. of Data points	Control Feed COD (mg/L)	Test Feed COD (mg/L)	Control Effluent COD (mg/L)	Test Effluent COD (mg/L)	% COD Reduced in Control	% COD Reduced in Test	% Increase in COD Reduction in Test
20-day	30	48,351	48, 351	26,500	26,877	45	44	-2
15-day	7	38,386	38,386	29,886	27,029	22	30	25
10-day	19	38,718	38,718	29,329	25,674	24	34	28
7-day	24	42, 167	42, 167	30,691	28,522	27	32	16

Table B-8. Total Particulate Solids Data Obtained from the Control and the Test Digesters.

HRT	No. of Data points	Control Feed TPS (mg/L)	Test Feed TPS (mg/L)	Control Effluent TPS (mg/L)	Test Effluent TPS (mg/L)	% TPSr in Control Digester	% TPSr in Test Digester	% Increase in TPS Reduction in Test
20 day	30	39,517	39,517	23,072	20,323	42	49	14
15 day	7	37,698	37,698	28,611	26,277	24	30	20
10 day	19	36,517	36,517	29,554	25,650	19	30	37
7 day	24	37,256	37,256	26,439	25,447	29	32	9

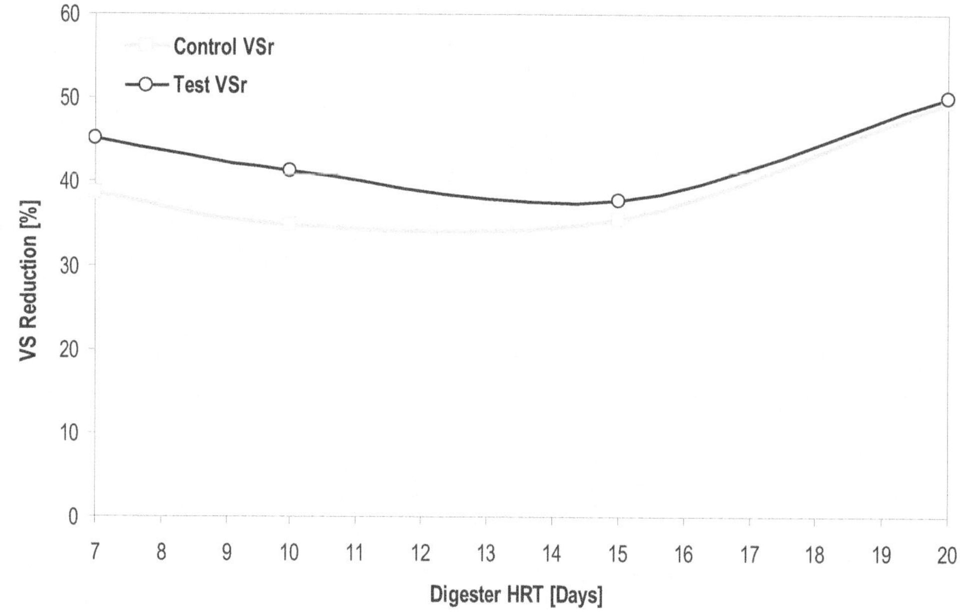

Figure B-13. Volatile Solids Reduction in Control and Test Digesters.

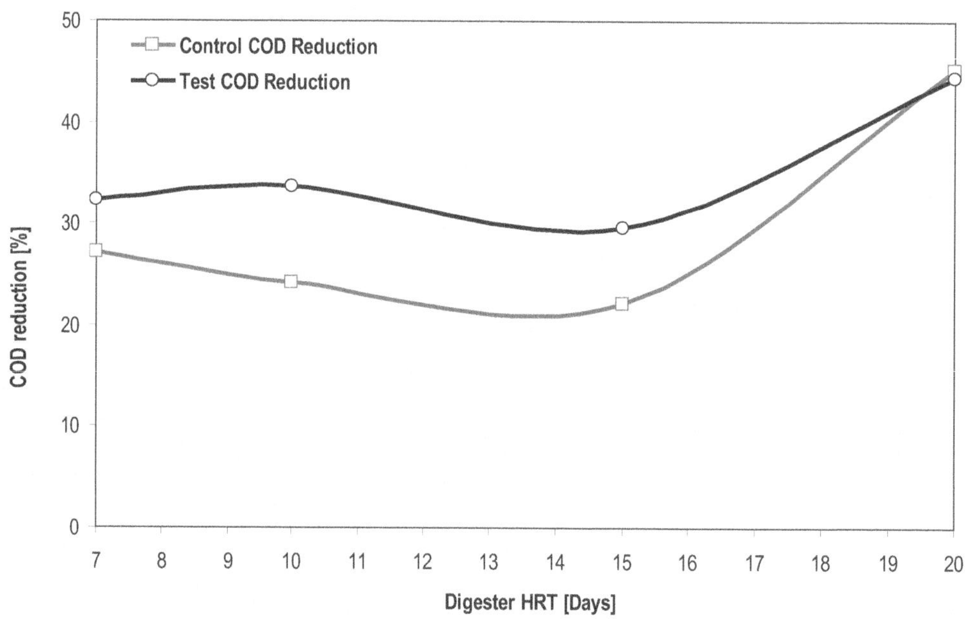

Figure B-14. COD Reduction in Control and Test Digesters.

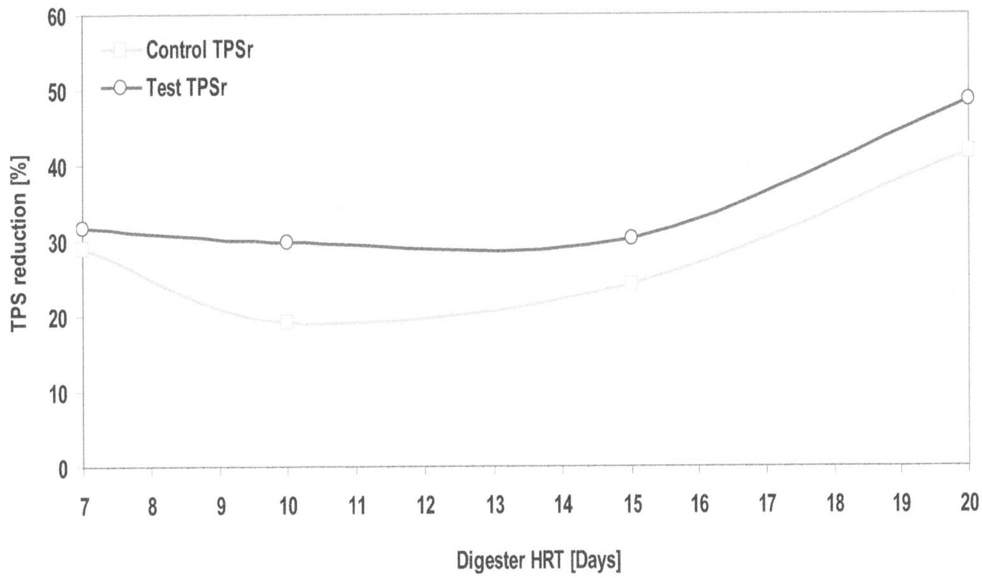

Figure B-15. Total Particulate Solids Reduction in Control and Test Digesters.

B.5.3.5 Biogas Production

Total biogas produced by the digesters was measured using gas meters as mentioned in the experimental summary section. Percent carbon dioxide (CO_2) was measured in some (but not in all) of the biogas samples, with the difference assumed to be methane. Based on this simplified testing approach, it was estimated that the control digester gas had an average of 29% CO_2 and 71% methane, while the test digester had an average of 26% CO_2, and 74% methane. These partial results would indicate a slight advantage in the caloric value of the biogas from the test digester. However, because the biogas composition data was collected infrequently, all evaluations in this report were done on the basis of total biogas and not methane production to avoid potentially overestimating the caloric value of the biogas. Further testing with more comprehensive biogas composition evaluations would be required to fully assess this issue.

Table B-9 summarizes the total volume of biogas produced at different digester HRTs. For all the HRTs tested, more biogas was produced in the test than in the control. This is interesting, because as discussed in the previous section, at a 20-day HRT, the test digester did not exhibit additional VSr or CODr when compared with the control digester. The amount of biogas generated is commonly related to the extent of VSr and CODr. Ruling out a laboratory or instrument measurement error, an increase in biogas production without the increase in VSr and CODr would likely be due to the differences in the actual nature of the volatile matter present in the untreated and MicroSludge™-pretreated sludge.

Table B-9. Total Volume of Biogas and Methane Produced over Testing Period of Various Experimental Phases.

HRT	No. of days of operation	Total Biogas Production in Control (L)	Total Biogas Production in Test (L)	% Increase in Biogas Production in Test
20	86	20,392	24,045	15
15	17	5,376	6,093	12
10	51	21,954	26,796	18
7	77	22,619	25,304	11

Table B-10 summarizes the biogas production rate of control and test digesters. Figure B-16 shows a definite trend of decreased biogas production rate as the HRT increases in both the control and test digesters. These results also indicate that at all HRTs, the rate of biogas generation of the MicroSludge™-treated TWAS was higher than the control unit, which would support the use of shorter detention time anaerobic digesters (10- to 15-day HRT) with significant capital cost savings as compared to units sized for 20-day HRT or above.

Table B-10. Biogas Production Rate from the Control and the Test Digesters.

HRT	No. of Data points	Total Gas Production in Control (L/hour)	Total Gas Production in Test (L/hour)	% Increase in Biogas Production in Test
20	30	11	13	16
15	7	13	15	13
10	19	19	21	10
7	24	26	30	13

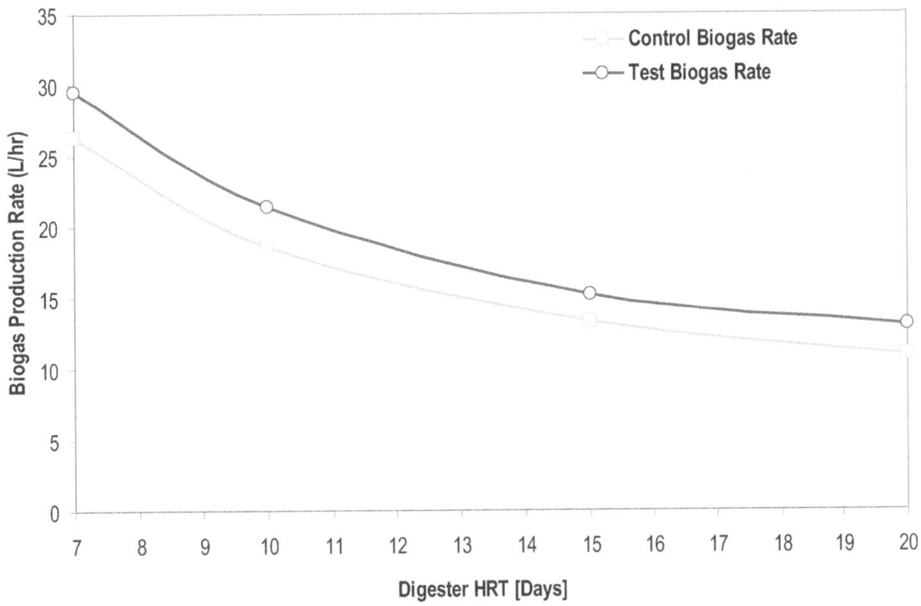

Figure B-16. Biogas Production Rate in Control and Test Digesters.

Tables B-11 and B-12 summarize the data organized in terms of biogas yield (biogas production per mass of VSr and CODr) from the control and the test digesters. VSr-based yields show a diminishing trend as HRTs are reduced. CODr trends, except for the unusual results reported for the 15-day HRT tests, are somewhat similar.

Table B-11. Comparison of Biogas Yield on a VSr Basis in Control and Test Digesters.

HRT	No. of Data points	Control Gas Yield (L biogas/ g VSr)	Test Gas Yield (L biogas/ g VSr)	Increase in Yield (%)
20	30	1.08	1.20	10
15	7	1.31	1.46	11
10	19	1.53	1.63	6
7	24	1.31	1.27	-3

Table B-12. Comparison of Biogas Yield on a CODr Basis in Control and Test Digesters.

HRT	No. of Data points	Control Gas Yield (L biogas / g CODr)	Test Gas Yield (L biogas / g CODr)	Increase in Biogas Yield (%)
20	30	0.81	0.90	9
15	7	1.70	3.00	43
10	19	1.47	1.41	-4
7	24	1.67	1.67	0.2

Research by McCarty (1972) demonstrated that anaerobic digestion of different substrates could result in varying biogas yields. Hence, the observed higher biogas yield at essentially equivalent VSr and CODr levels for test and control digesters operating at the 20-day HRT could be explained by a difference in the characteristics of the methanogenesis substrate, as supported by the significantly higher VFA concentrations observed in the MicroSludge™-treated samples. This higher level of substrate availability could also explain the higher methanogenic activity in the test digesters observed from the BESA incubation tests conducted at Virginia Tech and discussed in Section 5.2.2.4 of this document. An obvious next step in related research should focus on a more rigorous study of the overall issues involving substrate characteristics, biogas yields, and biomass yields.

B.5.3.6 Pathogenic Indicator Bacteria

Table B-13 compares the total and fecal coliform and *E. Coli* counts obtained from the control and the test sludge samples analyzed in Bucknell University.

Table B-13. Pathogenic Indicator Bacterial Count for Control and Test Digesters.

	Total Coliform (log MPN/g DS)	Fecal Coliform (log MPN/g DS)	*E. Coli* (log MPN/g DS)
Control Digester	6.87	5.82	5.82
Test Digester	6.88	5.46	5.41

Notes:
MPN/g = most probable number per gram
DS =dry solids

There were no statistically significant differences were found between the bacterial densities of the test and control digesters. This is likely because the high amounts of pathogenic indicator bacteria in the PS also fed to both digesters.

B.5.4 TWAS Only at 15 Days HRT

B.5.4.1Experimental Setup

The second pilot-scale digesters setup and operation was very similar to the first, except that the control digester and test digesters were fed TWAS only and that both digesters were operated at a 15-day HRT only. These tests were conducted between April 28 and August 7, 2009 (a total of 102 days of operation).

The analytics for these tests were essentially the same as that mentioned previously. Gas volumes in these digesters were measured using wet-tip flip-flop meters. These gas meters were calibrated based on the Ritter style wet-test gas meters used for the first set of digesters.

TWAS samples were microscopically examined by Dr. David Jenkins to evaluate the effect of the MicroSludge™ technology on filamentous organisms.

Samples of raw and anaerobically digested sludge from control and test digesters were sent to Dr. John Novak at Virginia Tech for organic sulfur odor testing, and batch digestion studies to determine the rates of change in VFA generation, soluble phosphorus production, and organic nitrogen speciation via 5-day incubation.

Samples of untreated and MicroSludge™ treated sludge samples were also sent Dr. Matthew Higgins at Bucknell University for total coliform, fecal coliform and *E. Coli* analysis using EPA method 1680.

B.5.4.2Results and Discussion

Microscopic Analysis

Two of the several micrographs provided by Dr. Jenkins of samples before and after MicroSludge™ pretreatment are shown in Figure B-17. The untreated WAS sample (Figure B-17a) contained firm, irregular, compact floc masses with some polyphosphate-accumulating organisms and some filamentous organisms. The filaments were mostly short and showed no effect on floc structure. They were morphologically identified to belong to Type 0041, Type 0675 and *Microthrix parvicella.* These filaments are all associated with low fed to microorganisms (F/M) ratio (high SRT). The MicroSludge™ pre-treated samples (Figure B-17b) contained a mass of dispersed bacteria, very few filament fragments and inorganic particles. MicroSludge™ seemed to have converted the flocs largely into dispersed bacteria and to have disrupted the filamentous microorganisms. These results support prior observations that foaming induced by filamentous organism could be abated by MicroSludge™ treatment of TWAS prior to digestion.

Figure B-17. (a) Untreated Sludge Sample Showing Filamentous Organisms, (b) MicroSludge™-Treated Sludge Showing Disrupted Bacteria with no Filamentous Organisms.

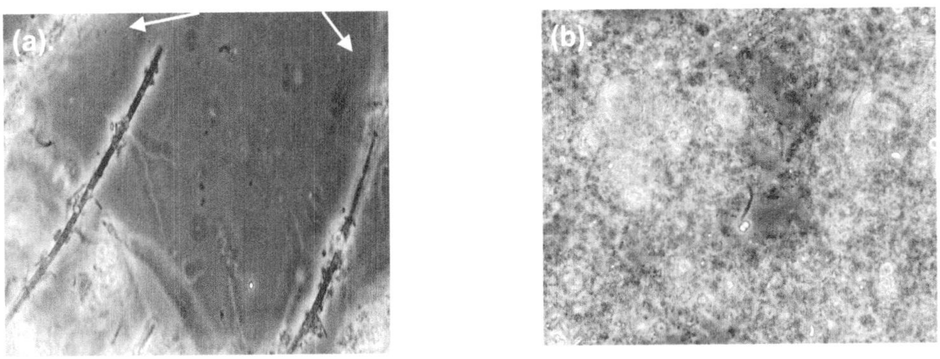

Organic Sulfur Odor Testing

Figure B-18 shows the TVOSC measured from sludge samples obtained from (a) control and (b) test digesters during 5 days of anaerobic incubation.

Figure B-18. (a) TVOSC for Untreated TWAS with and without BESA Added,
(b) TVOSC for Pretreated TWAS with and without BESA Added.

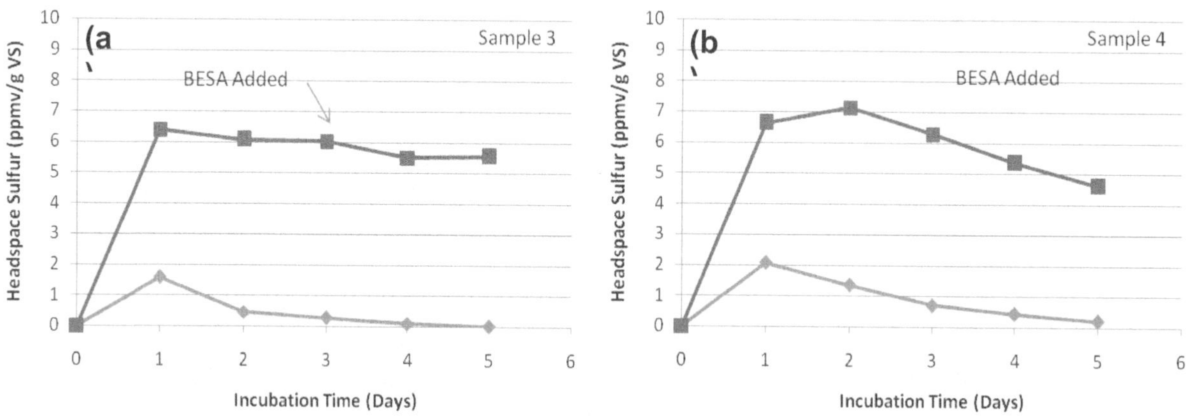

TVOSC generated from both untreated and MicroSludge™ pretreated TWAS samples were observed to be similar and generally very low as compared to the results presented in Figure B-15 for the tests that included PS in addition to TWAS. This would indicate that the primary source of sulfur-related odors is PS, and not the TWAS. BESA addition tests also indicated very little difference in the methanogenic activity of the anaerobically incubated sludge samples. This is in sharp contrast with the results from the first set of tests conducted with both PS and TWAS, in which significantly higher methanogenic activity was reported from samples of test digester. This could be explained by the likely assumption that that the source of both sulfur-related odor precursors and sulfur-reducing bacteria is primarily the PS. It should be noted that sources other than sulfur exist that would result in noxious odor releases as part of the anaerobic digestion process and that were not evaluated as part of this study.

Volatile Solids, COD, and TPS Reduction

Table B-14 summarizes the volatile solids data reported for the control and the test digesters. The volatile solids concentration of the feed to both the digesters was maintained same. The test digester showed an increase of 17% in VSr when compared to the control digester at a 15-day HRT. This result was consistent with the Chilliwack WWTP study, where at an HRT of 13 days, the test digester exhibited 18% increase in average VSr when compared with a control digester.

Table B-14. Volatile Solids Data Obtained from the Control and the Test Digester.

HRT	Control Feed VS (mg/L)	Test Feed VS (mg/L)	Control Effluent VS (mg/L)	Test Effluent VS (mg/L)	% VS Reduction in Control	% VS Reduction in Test	% Increase in VS Reduction in Test
15	29,929	29,929	19,893	17,800	34	41	17

Table B-15 shows the COD data obtained from the control and the test digesters. The COD concentration of the feed to both the digesters was maintained same. The test digester showed 23% more COD reduction than the control digester at 15 day HRT.

Table B-15. COD Data Obtained from the Control and the Test Digesters.

HRT	Control Feed COD (mg/L)	Test Feed COD (mg/L)	Control Effluent COD (mg/L)	Test Effluent COD (mg/L)	% COD Reduction in Control	% COD Reduction in Test	% Increase in COD Reduction in Test
15	44,776	44,776	32,746	29,468	27	35	23

Table B-16 shows the TPS data obtained from the control and the test digesters. The TPS concentration of the feed to both the digesters was maintained same. The test digester showed 30% more TPSr than the control digester at 15-day HRT.

Table B-16. Total Particulate Solids Data Obtained from the Control and the Test Digesters.

HRT	Control Feed TPS (mg/L)	Test Feed TPS (mg/L)	Control Effluent TPS (mg/L)	Test Effluent TPS (mg/L)	% TPS Reduction in Control	% TPS Reduction in Test	% Increase in TPS Reduction in Test
15	40,400	40,400	29,595	25,040	27	38	30

The difference of 17% in the VSr between control and test digesters fed only TWAS is considerably higher than the 5% observed for the co-digestion of PS and TWAS at the same 15-day digester HRT (refer to Table B-7. This could be explained by the significant presence of volatile solids in the PS that are not affected by MicroSludge™ treatment. A similar explanation could be drawn for TPSr. A 10% increase in TPSr was observed at 15-day HRT in TWAS only digester, when compared to co-digestion of PS and TWAS at the same 15-day digester HRT (refer to Table B-8).

On the other hand, the values of CODr are very similar at 23 and 25% for TWAS only and codigestion respectively. This could be explained by understanding that the basic mechanism behind MicroSludge™ treatment is the lysing of cell mass (relatively difficult to degrade suspended matter), and the subsequent exposure of additional COD now as readily biodegradable soluble matter to the digestion process. Again, no solubilization of suspended COD should be expected from the PS fraction of the feed as a result of MicroSludge™.

Table B-17 summarizes the biogas-related information reported for the TWAS only digestion testing period. This data shows a significant difference between control and test digesters total biogas production (50%), which is considerably greater than the values reported for VSr (17%) and CODr (25%), suggesting (similarly to the co-digestion test results) that the biogas yield may be improved by the MicroSludge™ pre-treatment of the TWAS. Again, these results would indicate that incorporating a MicroSludge™ process ahead of the mesophilic anaerobic digestion of TWAS would allow the use of shorter digester HRTs (thus reducing facility requirements); it would produce higher biogas quantities (higher co-generation potential); as well as lower quantities of digested sludge (lower biosolids processing and disposal/reuse requirements) when compared against digestion of untreated TWAS only.

Table B-17. Summary of Biogas Information for Digestion of TWAS only Testing Period.

Parameter (15-day Digester HRT)	Control	Test	Increase in Test, %
Total Biogas Production, (L)	7,884	15,506	50
Biogas Production Rate, (L/hour)	8	11	27
Biogas Yield (L gas/ g volatile solids reduced)	1.00	2.30	56
Biogas yield (L biogas / g COD reduced)	1.38	1.14	-21

Pathogenic Indicator Bacteria

Table B-18 compares the total and fecal coliform and *E. Coli* counts obtained from the analysis of untreated TWAS and MicroSludge™-treated TWAS by Bucknell University.

Table B-18. Pathogenic Indicator Bacterial Count for Control and Test TWAS Samples.

Digester	Total Coliform (log MPN/g DS)	Fecal Coliform (log MPN/g DS)	*E. Coli* (log MPN/g DS)
Control	7.76	6.16	6.23
Test	6.18	5.24	5.32

These results show a slight decrease (less than a log reduction) in pathogenic indicator organism as a result of treatment of the TWAS samples with MicroSludge™. Based on these results, it can be concluded that MicroSludge™ treatment does not have a significant impact on pathogen reduction ahead of the anaerobic digestion process.

B.6 Conclusions

The independent evaluation of the data collected by Paradigm Environmental Technologies as part of the demonstration study conducted at the Des Moines WRF provided an excellent opportunity to assess the performance of this WAS conditioning technology as applied using a full-scale installation of the MicroSludge™ process. Although the tests conducted using full-scale anaerobic digesters were difficult to control and thus made impossible the determination of potential benefits from pre-treating TWAS before digestion, it did offer the opportunity to evaluate the operational reliability and maintenance requirements of a commercial scale MicroSludge™ system. The subsequent pilot scale digester tests (which relied on feed from the full-scale MicroSludge™ process), allowed the actual determination of process performance differences under much more controlled conditions. The main conclusions that can be derived from this study include the following:

B.6.1 MicroSludge™ Full-Scale Maintenance and Reliability Assessment

Results from the demonstration study indicated that the full-scale system can be designed to provide a very high level of operational reliability, provided the following measures are included as part of a design:

- Running a scheduled maintenance program to replace wear components before failure
- Incorporating redundant screens, macerator and cell disruptor (CD) feed pump, all on hot standby
- Minimizing CD feed pressure to reduce pump rotation speed and correspondingly decrease wear
- Optimized design of CD valve and valve block
- Optionally including hydrocyclones to remove grit from WAS

Related to the first item in the preceding list, Table B-19 presents the expected maintenance requirements for a full-scale installation of similar processing capacity as the one tested at Des Moines (approximately 96 m^3/day or 26,000 gal/day).

Table B-19. Expected Maintenance Schedule for a Full-scale MicroSludge™ System.

Component	Category	Period (weeks)	Labor (hours)	Total Annual Labor (hours)
Cylinders	Packing	4	2	26
Cylinders	Gland	52	1	1
Check Valves	Balls	8	2	13
Check Valves	Valve Seats	52	2	2
Homogenizing Valves	Replacement	4	1	13
Homogenizing Valves	Block	52	2	2
Filters	Cleaning	52	8	8
Pumps	Lobe and Wear plate	26	2	4
Cell Disrupter	Lubrication	26	4	8
General Maintenance	Cleaning and Lubrication	52	8	8

B.6.2 Co-Digested PS and TWAS Operating at Varying HRTs

- Organic sulfur emissions from MicroSludge™-treated sludge were about one-third of that of the control digester, indicating lower odor generation potential for MicroSludge™-treated sludge.
- Anaerobic incubation batch testing incorporating bromoethane sulfonic acid (BESA) addition identified higher methanogenic activity for MicroSludge™ treated TWAS.
- Batch incubation testing also indicated that MicroSludge™ pre-treatment can lead to the direct formation of VFAs due to solubilization of sludge samples. The presence of significantly higher amounts of VFAs will consequently reduce the reaction time required for anaerobic digestion, since hydrolysis and acidogenisis (the necessary preceding steps before methanogenesis) have already occurred. By comparison, an acid-phase anaerobic digestion system is typically designed for a 2-day SRT; hence, this data suggests that the rate of digestion is significantly increased by the MicroSludge™ process.

- Increase in the nutrient content of digested sludge reject streams (such as those generated from subsequent dewatering steps) are one of the negative impacts normally associated with the pre-treatment of TWAS ahead of anaerobic digestion (Sandino et al., 2008). However, MicroSludge™ treatment of TWAS at the Des Moines WRF did not appear to increase the levels of soluble phosphorus and N_2 in the digester content, albeit the higher VSr levels achieved at the shorter digester HRTs. This probably was the result of formation of calcium phosphate precipitate and stripping of ammonia at higher pH levels caused by the caustic addition step that is part of the conditioning process.
- VSr and CODr results documented at the Des Moines WRF appear to indicate that MicroSludge™ technology may not be of much advantage to anaerobic digesters operating at or above 20 days HRT, and that its potential to increase the extent of biomass reduction is only achievable at lower HRTs. This result is consistent with previous studies at other facilities and suggests that MicroSludge™ is capable of increasing the rate of digestion and, therefore, may reduce the retention time (thus digester volume) required for anaerobic digestion. Also, higher levels of VSr reduction at the lower digester HRTs would result in smaller quantities of digested biosolids requiring further in-plant processing and ultimate disposal/reuse outside the treatment facility.
- Cumulative gas production was higher in the test unit than in the control under all the evaluated digester HRTs, even though results also showed that at a 20-day HRT there was no significant difference in VSr and CODr levels. An increase in biogas yield at essentially equivalent VSr and CODr levels could be explained by a difference in the characteristics of the methanogenesis substrate, as supported by the significantly higher VFA concentrations, and/or higher methanogenic activity of the MicroSludge™ pretreated sludge. Higher biogas production capabilities would increase the co-generation potential at a treatment facility.
- Results from Des Moines pointed to a definite trend of increased biogas production rate as the HRT was decreased in both the control and test digesters, with even higher rates for the MicroSludge™ treated TWAS. Again, this finding which would support the use of shorter detention time anaerobic digesters (~ 10-15 day HRT) with significant capital and O&M cost savings as compared to units sized for 20 d HRT or above.
- There were no statistically significant differences between the bacterial densities of the test and control digesters.

B.6.3 TWAS Only at 15 Days HRT

- Microscopic evaluations conducted indicate that MicroSludge™ processing of TWAS destroyed filamentous organisms and thus would result in lower filamentous growth-related foaming incidents in anaerobic digesters.
- Sulfur-related emissions generated from both untreated and MicroSludge™-pre-treated TWAS samples were observed to be similar and generally very low compared to the results obtained with co-digestion of sludges. This indicates that the primary source of sulfur-related odors at the Des Moines WRF was PS, and not the TWAS. Batch digestion tests also indicated very little difference in the methanogenic activity of the anaerobically incubated TWAS samples. This is in sharp contrast with the results from the co-digestion tests, in which significantly higher methanogenic activity was reported from samples of the MicroSludge™-treated digester. This difference could be explained if the PS is considered to

be the source of both sulfur-related odor precursors and sulfur-reducing bacteria. It should be noted that there are sources other than sulfur that would result in noxious odor releases as part of the anaerobic digestion process and that were not evaluated as part of this study.

- Tests indicated a much larger difference in the VSr between control and test digesters fed TWAS-only when compared to the results from co-digestion at the same 15-day digester HRT. This could be explained by the significant presence of volatile solids in the PS that are not affected by MicroSludge™ treatment. On the other hand, the values of CODr observed at the Des Moines WRF were very similar to those observed in the co-digestion tests. This could be explained by understanding that the basic mechanism behind MicroSludge™ treatment is the lysing of cell mass which is relatively difficult to degrade suspended matter, the subsequent production of high levels of soluble COD (mainly VFAs), and from the fact that the organic matter in PS is largely unaffected by this sludge conditioning process.

- Biogas data showed a significant difference between control and test digesters total biogas production, which is considerably greater than the values reported for VSr and CODr. This suggests (similarly to de co-digestion test results) that the biogas yield may be improved by the MicroSludge™ pre-treatment of the TWAS. Again, these results would indicate that incorporating a MicroSludge™ process ahead of an mesophilic anaerobic digester for TWAS would allow the use of shorter HRTs (thus reducing facility requirements); it would produce higher biogas quantities (higher co-generation potential), as well as reducing the quantities of digested sludge and thus lowering the requirements for biosolids processing and disposal/reuse when compared against digestion of untreated TWAS only.

- Results at the Des Moines WRF showed less than a log reduction in pathogenic indicator organism as a result of treatment of the TWAS samples with MicroSludge™. Based on these results, it can be concluded that MicroSludge™ treatment did not have a significant impact on pathogen reduction ahead of the anaerobic digestion process.

B.7 References

APHA (1998) Standard Methods for the Examination of Water and Wastewater. 20th edition. American Public Health Association. Washington, D.C.

McCarty (1972) "*Energetics of Organic Matter Destruction," Water Pollution Microbiology*. pp. 91-118. John Wiley. New York.

Rabinowitz, B. and R., Stephenson (2005) *Improving Anaerobic Digester Efficiency by Homogenization of Waste Activated Sludge.* Proceedings, WEFTEC 79th Annual Technical Exhibition and Conference. Washington, D.C. November 1.

Sandino, S., L. Fillmore, D. Whitlock, and R. Roxburgh (2008) *Current State of the Practice of Sludge Reduction Technologies.* Proceedings from the 2008 Water Environment Federation Residuals and Biosolids Specialty Conference.

Stephenson, R.J. and H. Dhaliwal (2000) *Method of Liquefying Microorganisms Derived fromWastewater Treatment Processes.* U.S. Patent No. 6,013,183.

Stephenson, R., S. Laliberte, P. Hoy, and A. Drew (2009) *Demonstration of Faster Digestion and Residual Reduction with MicroSludge™ Pre-treatment of Waste Activated Sludge at Des Moines Wastewater Reclamation Facility.* Proceedings, WEA of Ontario. September 13-15.

APPENDIX C

POTW QUESTIONNAIRE

C.1 Facility Information

- Contact Name:
- Plant Name:
- Address:
- State, Province:
- Country:
- Phone:
- Email:

C.2 Questions

C.2.1 General

- What type of solids reduction technology does your plant have?
- Where in the plant is the technology located?
- How long has this equipment been in operation?
- What is the plant annual average influent flow and loading rate?
- Please provide solids mass balance data, before and after the technology was installed.
- Can you cite other benefits associated with the technology, for example:
  - Has the technology been helpful in abating foam?
  - Has the technology been helpful in abating odor?
  - Has the technology been helpful in abating filamentous bacteria? Do you have SVI or other data to support this?
  - Has the technology been helpful in augmenting nutrient removal in the liquids process? Do you have nutrient data to support this?

C.3 Operation

- What are the labor hours on a monthly basis?
- What are the chemical quantities and costs for this technology?
- What types of chemicals are used in conjunction with this technology?
- What quantities of chemicals are used on a monthly basis?
- What are the chemical costs on a monthly basis?
- What is the power consumption of the technology on a monthly basis?
- What are the power costs of the technology on a monthly basis?

- Please provide any operational issues associated with the equipment?

C.4 Maintenance

- Have you had any maintenance issues?
- How many FTEs do you spend on maintenance?
  - Selected project individual costs

APPENDIX D

COMPREHENSIVE LIST OF PROCESSES

Manufacturer	Process	In Operation	Design Flow (MLD)	Municipal/Industrial	Range of Reported Impacts*	Op Ins
ogical Treatment Processes						
A	Sequenced Wastewater Treatment	1998	0.013	Municipal	0.2 kg TSS/kg CODr; 0.09 kg VSS/kg CODr	1
c	Sequenced Wastewater Treatment	2003		Municipal		
c	Sequenced Wastewater Treatment	Mar-03 to Jun-04	1.14	Municipal		
c	Sequenced Wastewater Treatment	2003		Municipal	Up to 39% > VSR	1
v Environmental	Sequenced Wastewater Treatment	2007?			?85% VSS Reduction	1 In C
ndi	Sequenced Wastewater Treatment			Synthetic	58% less WAS	Bench
X AB	Sequenced Wastewater Treatment			Municipal	14% less WAS	Large
X AB	Sequenced Wastewater Treatment			Industrial - pulp and paper	80% less WAS	Full-S
X AB	Sequenced Wastewater Treatment			1 Municipal; 1 Industrial	14% to 80% less WAS	1 Ope

Manufacturer	Process	In Operation	Design Flow (MLD)	Municipal/Industrial	Range of Reported Impacts*	No. of Operational Installations
Kuraray	Fixed-Film (PVA-Gel Beads) with TOX	Mar-01 (Type C), switched to Type S Jan-07	400 m3/d	Industrial - Formalin (1,500 mg/L) + PVA (100 mg/L), BOD = 2,000mg/L, TSS = 0	>90%	Operational
Kuraray	Fixed-Film (PVA-Gel Beads) with TOX	February 2001 (Type C, switched to Type M on April 2006)	20 m3/d	Industrial - Methanol (90%) + TBA. BOD 1000 - 4,000 mg/L, av 2,500 mg/L, TSS = 0	>90%	Operational
Kuraray	Fixed-Film (PVA-Gel Beads) with TOX	May-07	6.813	Industrial - Dye chemicals and PVA - BOD 250 mg/L, TSS After Screening 60 mg/L	70% Sludge Reduction	Operational
Kuraray	Fixed-Film (PVA-Gel Beads) with TOX			Municipal	>90% After Primary Sedimentation	Pilot (200 L Vessels)
Kuraray	Biological Cell Lysis	2001	0.02 - 6.8		70 to 100%	3 Industrial; 1 Municipal Large Bench-Scale
Premier Wastewater International (PWI)	Enhanced Solids Reduction with Multi-Action Conditioner	2002?	0.3785	Municipal	90% less WAS	Demonstration Plant
PWI	ESR with MAC		0.662375	Industrial - Juice Concentrate Plant		
Premier Wastewater International (PWI)	Enhanced Solids Reduction with Multi-Action Conditioner					
United Utilities/Monsal	Enzymatic Hydrolysis	2002	12	Municipal	59% VSR	
United Utilities/Monsal	Enzymatic Hydrolysis	2005	24	Municipal	62.5% VSR	

Manufacturer	Process	In Operation	Design Flow (MLD)	Municipal/Industrial	Range of Reported Impacts*	No. of Operational Installations
United Utilities/Monsal	Enzymatic Hydrolysis	2005	30	Municipal	54.3% VSR	
United Utilities/Monsal	Enzymatic Hydrolysis	2006	54	Municipal	55% VSR	
United Utilities/Monsal	Enhanced/ Enzymatic Hydrolysis	2002	12-54	Municipal	22% > VSR	4 Operational, 5 Construction
Siemens	Cannibal	Winter 2004	7.57	Municipal		Operational
Siemens	Cannibal	Winter 2004	2.84	Municipal		Operational
Siemens	Cannibal	Summer 2005	22.71	Municipal		Operational
Siemens	Cannibal	Spring 2005	5.68	Municipal		Operational
Siemens	Cannibal	Fall 2006	60.56	Municipal		Operational
Siemens	Cannibal	Summer 2007	6.62	Municipal		Operational
Siemens	Cannibal	Spring 2006	5.68	Municipal		Operational
Siemens	Cannibal	Spring 2006	37.85	Municipal		Operational
Siemens	Cannibal	Fall 2007	28.39	Municipal		Operational
Siemens	Cannibal	Fall 2007	0.30	Industrial (Cheese Factory)		Operational
Siemens	Cannibal	Summer 2007	15.90	Municipal		Operational
Siemens	Cannibal	Summer 2007	20.82	Municipal		Operational
Siemens	Cannibal	Summer 2007	17.03	Municipal		Operational
Siemens	Cannibal	Fall 2007	11.36	Municipal		Operational
Siemens	Cannibal	Summer 1998	2.84	Municipal		Operational
Siemens	Cannibal	Fall 1999	1.70	Municipal		Operational
Siemens	Cannibal	Summer 2005	22.71	Municipal		Operational

Manufacturer	Process	In Operation	Design Flow (MLD)	Municipal/Industrial	Range of Reported Impacts*	No. of Operational Installations
Siemens	Cannibal	NA	13.63	Municipal		In Construction
Siemens	Cannibal	NA	6.06	Municipal		In Construction
Siemens	Cannibal	NA	60.56	Municipal		In Construction
Siemens	Cannibal	NA	1.14	Municipal		In Construction
Siemens	Cannibal	NA	30.28	Municipal		In Construction
Siemens	Cannibal	NA	22.71	Municipal		In Construction
Siemens	Cannibal	NA	12.11	Municipal		In Construction
Siemens	Cannibal	NA	2.84	Municipal		In Construction
Siemens	Cannibal	NA	15.14	Municipal		In Construction
Siemens	Cannibal	NA	10.60	Municipal		In Construction
Siemens	Cannibal	NA	5.68	Municipal		In Construction
Siemens Cannibal	Combination Processes	1999	0.3–60.6	1 Industrial (Cheese Factory)		17 Operational; 11 Construction
	Combination Processes					
	Combination Processes					
	Combination Processes	2008	56.8	Municipal		
CATABOL	Combination Processes					1 Operational, 1 Pilot, 1 Construction
Aquarius Technologies	Combination Processes		0.3	Industrial		Operational
Aquarius Technologies	Combination Processes		0.795	Municipal		Operational
Aquarius Technologies	Combination Processes		0.15	Municipal		Operational

Manufacturer	Process	In Operation	Design Flow (MLD)	Municipal/Industrial	Range of Reported Impacts*	No. of Operational Installations
Aquarius Technologies	Combination Processes		0.15–0.8	1 Industrial; 2 Municipal	No Biological Sludge	3 Operational
CHEMICAL TREATMENT PROCESSES						
Praxair	Ozonation	May-06			35% to 40% Sludge Reduction	Pilot
Praxair	Ozonation	Jul-07	26.495		Maximum 80% Sludge Reduction	Operational
Praxair	Ozonation	May-06	26.495	Municipal with 60% textile	35% to 80%	1 Operational
Degremont	Ozonation					Pilot
Degremont	Ozonation					Pilot
Degremont	Ozonation					Pilot
Degremont	Ozonation	Planned for 2007				Pilot
Degremont	Ozonation	4–12 Months				3 pilot
University - Horng	UASB + Chemical Hydrolysis			Industrial - Maleic Dehydrate	90% less WAS	1 Operational
Heinzman & Engel	Thermal + Chemical Hydrolysis			Municipal	17% less WAS	Pilot on 1 Bioreactor of 8
Heinzman & Engel	Thermal + Chemical Hydrolysis			Municipal		Full-scale digester
Heinzman & Engel	Thermal + Chemical Hydrolysis			Municipal		Full-Scale Digester
Heinzman & Engel	Thermal + Chemical Hydrolysis			Municipal		Full-Scale Digester

Manufacturer	Process	In Operation	Design Flow (MLD)	Municipal/Industrial	Range of Reported Impacts*	No. of Operational Installations
Heinzman & Engel	Thermal + Chemical Hydrolysis			Municipal		3 Operational
PMC Biotec	Chem/Thermophilic Digestion			Groundwater with Organics	>80%	Operational
PMC Biotec	Chem/Thermophilic Digestion			Splenda Sugar Waste	>80%	Operational
PMC Biotec	Chem/Thermophilic Digestion	2000	57 m^3/d	Toluene and Exotics	>80%	Operational
PMC Biotec	Chem/Thermophilic Digestion			Pharmaceutical	>80%	Operational
PMC Biotec	Chem/Thermophilic Digestion	2002	24.6025	pharmaceutical	>80%	Operational
PMC Biotec	Chemical Lyses	Before 2000		Industrial - v. High-Strength, to Sewer	>80%	5
PHYSICAL TREATMENT PROCESSES						
Ultrawaves	Sonication	1997–1999	16.99	Municipal	• Reduction in digestion time from 20 to 4 days without losses in degradation efficiency • Increase in biogas production by a factor of 4 (renewable energy!) • Reduction of digested sludge mass of 25%	First fundamental study on pilot scale by Technical University of Hamburg-Harburg, 3 years, 1997– 999

Manufacturer	Process	In Operation	Design Flow (MLD)	Municipal/Industrial	Range of Reported Impacts*	No. of Operational Installations
Ultrawaves	Sonication	1999 (6 Months)	9.99	Municipal	• Increase in VS destruction of 20% • Increase in biogas production of 20%	Preliminary test on pilot-scale by Technical University of Hamburg-Harburg, 6 months, 1999
Ultrawaves	Sonication	Jun-04	45.99	Municipal	• Increase in VS destruction of 30% • Increase in biogas production of 30% • Avoided the construction of a new anaerobic digester (3,000 m vol.)	1) Preliminary full-scale test, 4 months, 2002 2) Full-scale installation since Jun-04
Ultrawaves	Sonication	2003, 4 Months	26.00	Municipal	• Increase in biogas production of 15% • Improved sludge dewatering of 10%	Fundamental full-scale study by University of Armed Forces, Munich, 4 months, 2003
Ultrawaves	Sonication	Dec-04	4.01	Municipal	• Increase in VS destruction of 25% • Increase in biogas production of 25% • No foam or filamentous organisms present in the anaerobic sludge digester	1) Preliminary full-scale test, 3 months, 2004 2) Full-scale installation since December 2004 BNR oxidation ditch plant
Ultrawaves	Sonication	2004, 3 Months	12.98	Municipal	• Increase in VS destruction of 15% • Increase in biogas production of 25%	Full-scale test, 3 months, 2004

Manufacturer	Process	In Operation	Design Flow (MLD)	Municipal/Industrial	Range of Reported Impacts*	No. of Operational Installations
Ultrawaves	Sonication	2004–2005 (3 Months)	9.99	Municipal	• Increase in VS destruction of 25% • Increase in biogas production of 25%	Full-scale test, 3 months, 2004 – 2005
Ultrawaves	Sonication	2005–2005 (3 Months)	14.00	Municipal	• Increase in VS destruction of 15% • Increase in biogas production of 25%	Full-scale test, 3 months, 2004 – 2005
Ultrawaves	Sonication	May-05	14.99	Municipal	• Increase in VS destruction of 25% • Increase in biogas production of 25%	Full-scale installation since May 2005
Ultrawaves	Sonication	Jul-06	7.99	Municipal	• Increase in VS destruction of 25% • Increase in biogas production of 25%	Full-scale installation since July 2006
Ultrawaves	Sonication	Sep-06	12.98	Municipal		Full-scale installation since September 2006
Ultrawaves	Sonication	2006, 6 Months	5.00	Municipal	• Increase in VS destruction of 35% • Increase in biogas production of 35%	Full-scale test, 6 months, 2006
Ultrawaves	Sonication	Sep-06	11.01	Municipal		Full-scale installation since September 2006
Ultrawaves	Sonication	Sep-06	110.14	Municipal	• Increase in VS destruction of 15% • Increase in biogas production of 15%	Full-scale installation since September 2006

Manufacturer	Process	In Operation	Design Flow (MLD)	Municipal/Industrial	Range of Reported Impacts*	No. of Operational Installations
Ultrawaves	Sonication	Sep-06	88.19	Municipal		Full-scale installation since September 2006
Ultrawaves	Sonication	Mar-06	26.00	Municipal		Full-scale installation since March 2006
Ultrawaves	Sonication	Mar-06	43.98	Municipal	• Increase in sludge reduction of 15% • Increase in biogas production of 35% • Decrease in polymer consumption of 20%	Full-scale installation since March 2006
Ultrawaves	Sonication	Jan-2007 test	14.00	Municipal		Full-scale test, since January 2007
Ultrawaves	Sonication	Jul-07	50.00	Municipal		Full-scale installation in July 2007
Ultrawaves	Sonication	Jul-03	10.41	Municipal	• Sonication of return sludge • No foam and floating sludge in the aeration tank • Reduction of waste activated sludge production up to 20%	1) Preliminary full-scale test, 3 months, 2002 2) Full-scale installation since July 2003
Ultrawaves	Sonication	May-03		Municipal	• Sonication of return sludge • Reduction of waste activated sludge production up to 60%	1) Preliminary full-scale test, 3 months, 2002 2) Full-scale installation since May 2003

Manufacturer	Process	In Operation	Design Flow (MLD)	Municipal/Industrial	Range of Reported Impacts*	No. of Operational Installations
Ultrawaves	Sonication	2003, 3 Months	1.40	Municipal	• Sonication of return sludge • Reduction of waste activated sludge production of 20%	Preliminary full-scale test, 3 months, 2003
Ultrawaves	Sonication	Oct-04		Municipal	• Sonication of return sludge • Reduction of waste activated sludge production up to 60%	Full-scale installation since October 2004
Ultrawaves	Sonication	Oct-04		Municipal	• Sonication of return sludge • Reduction of waste activated sludge production up to 60%	Full-scale installation since October 2004
Ultrawaves	Sonication	2004/2005 3 Months	12.00	Municipal	• Sonication of return sludge • Reduction of waste activated sludge production of 25%	Preliminary full-scale test, 3 months, 2004 – 2005
Ultrawaves	Sonication	Mar-06	40.01	Municipal	• Sonication of return sludge • Reduction of waste activated sludge production of 25%	Full-scale installation since March 2006
Ultrawaves	Sonication	Jun-06	110.03	Municipal	• Sonication of return sludge • Reduction of waste activated sludge production of 15%	Full-scale installation since June 2006

Manufacturer	Process	In Operation	Design Flow (MLD)	Municipal/Industrial	Range of Reported Impacts*	No. of Operational Installations
Ultrawaves	Sonication	Mar-07	21.99	Municipal		Full-scale installation in March 2007
Ultrawaves	Sonication	Mar-07	7.00	Municipal		Full-scale installation in March 2007
Ultrawaves	Sonication	2002, 6 Months	2.61	Municipal	• Reduction in sludge volume index SVI from 110 to 60 mL/g • No foam and no filamentous organisms in the sludge digester	First fundamental study on pilot scale by Technical University of Hamburg-Harburg, 6 months, 2002
Ultrawaves	Sonication	2002, 6 Months	2.61	Municipal	• Reduction in sludge volume index SVI from140 to 55 mL/g • No foam and no filamentous organisms in the sludge digester	Fundamental study on pilot scale by Technical University of Hamburg-Harburg, 6 months, 2002
Ultrawaves	Sonication	2003, 3 Months	9.99	Municipal	• Reduction in sludge volume index SVI from150 to 75 mL/g • 30% reduction in foaming potential of digested sludge	Fundamental study on pilot scale by Technical University of Hamburg-Harburg, 3 months, 2003

Manufacturer	Process	In Operation	Design Flow (MLD)	Municipal/Industrial	Range of Reported Impacts*	No. of Operational Installations
Ultrawaves	Sonication	2003, 5 Months	5.00	Municipal	• Reduction in sludge volume index SVI from 120 to 80 mL/g • No floating sludge in the aeration tank	Preliminary full-scale test, 5 months, 2003
Ultrawaves	Sonication	2003, 3 Months	59.99	Municipal	• Reduction in sludge volume index SVI • 50% reduction in foaming potential of digested sludge	Fundamental study on pilot scale by University of Rostock, 3 months, 2003
Ultrawaves	Sonication	2005, 3 Months	48.98	Municipal	• Reduction in sludge volume index SVI from 150 to 60 mL/g • No floating sludge in the aeration tank	Preliminary full-scale test, 3 months, 2005
Ultrawaves	Sonication	2003, 3 Months	2.61	Municipal	• Reduction of waste activated sludge production up to 50% • Small improvement in denitrification by providing a carbon source	First fundamental study on pilot scale, 3 months, 2003
Ultrawaves	Sonication	Sep-06	10.79	Municipal	• Reduction of waste activated sludge production of 25% • Significant improvement in denitrification by providing a carbon source	1) Preliminary full-scale test, 5 months, 2006 2) Full-scale installation since September 2006
Ultrawaves	Sonication	May-03	1.4-110.14	Municipal WWTPs	15-35% >VSR & 0 - 10% >Dewatering; 15–60% less WAS	20 operational; 17 pilot

Manufacturer	Process	In Operation	Design Flow (MLD)	Municipal/Industrial	Range of Reported Impacts*	No. of Operational Installations
Sonico	Sonication	Mar-03	227.1	Municipal	-	Pilot
Sonico	Sonication	2000 to 2002.	83.27	Municipal	20% VSR Improvement	Pilot
Sonico	Sonication	Oct-03		Municipal	Equipment Problems	Pilot
Sonico	Sonication	Test Unit Apr-03		Municipal	Insufficient Time of Operation	Pilot
Sonico	Sonication	Test Jul-02–Jul-03	30.01505	Municipal	30% Guarantee	Pilot
Sonico	Sonication	Dec-02		Municipal	-	Operational
Sonico	Sonication	Jun-03–Nov-03	18.0166	Municipal	20% VSR Improvement	Pilot
Sonico	Sonication				30% WAS Reduction	Pilot
Sonico	Sonication	Oct-02	23.9969	Municipal	20% VSR Improvement	Operational
Sonico	Sonication	Oct-03		Municipal	-	Operational
Sonico	Sonication	Feb-04	36	Municipal	-	Operational
Sonico	Sonication	Apr-03		Municipal	-	Operational
Sonico	Sonication	2006	529.90	Municipal	Confidential	Operational
Sonico	Sonication	Jun-03		Municipal	-	Operational
Sonico	Sonication	2002	18–530	Municipal	VSR/WAS redn >20 to 30%; Dewaterability 0 to 3 Percentage Points	7 Operational; 7 Pilot
VTA	Sonication	Aug-05	5.58	Municipal		Operational
VTA	Sonication	Jul-05	4.20	Municipal		Operational
VTA	Sonication	Dec-04	36.00	Municipal		Operational
VTA	Sonication	Nov-04	28.80	Municipal		Operational

Manufacturer	Process	In Operation	Design Flow (MLD)	Municipal/Industrial	Range of Reported Impacts*	No. of Operational Installations
VTA	Sonication	Oct-03	11.40	Municipal		Operational
VTA	Sonication	Oct-03	5.40	Municipal		Operational
VTA	Sonication	Nov-03	24.00	Municipal		Operational
VTA	Sonication	Nov-03	4.80	Municipal		Operational
VTA	Sonication	from Oct-02 to Sep-03	9.12	Municipal		Pilot
VTA	Sonication	Nov-05	6.00	Municipal		Operational
VTA	Sonication	Oct-03	4.2–36	Municipal	Up to 25% >VSR	9 Operational; 1 Pilot
CAMBI	Thermal Hydrolysis	1996		Municipal	Dewaterability at 26%–30% (Belt-presses). No prior digestion.	Operational
CAMBI	Thermal Hydrolysis	1999		Municipal	No dewatering. Dewatering trials in 2001 gave 32% DS (belt-press)	Operational
CAMBI	Thermal Hydrolysis	2000		Industrial	No Dewatering	Operational
CAMBI	Thermal Hydrolysis	2000		Municipal	Solids reduction by 50%. Dewaterability increased from 18% to 29% (belt-presses)	Operational
CAMBI	Thermal Hydrolysis	2001		Municipal	Dewaterability at approx. 32% (belt-presses). No prior digestion.	Operational
CAMBI	Thermal Hydrolysis	2001		Municipal	Centrifuge dewatering. No polymers used. Mixed with yard waste to produce compost.	Operational

Manufacturer	Process	In Operation	Design Flow (MLD)	Municipal/Industrial	Range of Reported Impacts*	No. of Operational Installations
CAMBI	Thermal Hydrolysis	2002		Municipal	Dewaterability at 30% using Westfalia centrifuges. No prior digestion.	Operational
CAMBI	Thermal Hydrolysis	2002		Municipal and Industrial	Solids reduction by 43%. Dewaterability increased from 24% to 30% (centrifuges)	Operational
CAMBI	Thermal Hydrolysis	2002		Municipal		Pilot
CAMBI	Thermal Hydrolysis	2005		Municipal	Solids reduction by 50% (from 20,000 metric tpy to 10,000 tpy). Dewaterability from 20% to 31–33% (centrifuges)	Operational
CAMBI	Thermal Hydrolysis	2007		Municipal	28% DS with centrifuges during commissioning before optimization	Operational
CAMBI	Thermal Hydrolysis	2007		Municipal	They use a plate and frame press for final dewatering with 35% – 40% solids. New build.	In Construction
CAMBI	Thermal Hydrolysis	2007		Municipal	Test indicate increase from 19% to 27% DS in final dewatering	In Construction
CAMBI	Thermal Hydrolysis	2007				In Construction
CAMBI	Thermal Hydrolysis	2008		Municipal		In Construction
CAMBI	Thermal Hydrolysis	2008		Municipal		In Construction
CAMBI	Thermal Hydrolysis	2007		Municipal		In Construction

Manufacturer	Process	In Operation	Design Flow (MLD)	Municipal/Industrial	Range of Reported Impacts*	No. of Operational Installations
CAMBI	Thermal Hydrolysis	1996		1 Industrial; 1 Combined	VSR >40 – 50%; dewaterability > 6 to 11 percent points	10 Operational; 1 Pilot; 6 Construction
Veolia	BioThelys Thermal Hydrolysis	Aug-06	7.2	Municipal		
Veolia	BioThelys Thermal Hydrolysis	2006	4.56	Municipal		
Veolia	BioThelys Thermal Hydrolysis	2006	4.6–7.2	Municipal		2 Operational
Microsludge	Homogenization	May–Jul-04	17.5	Municipal	Increase VSR destruction; increase biogas production	
Microsludge	Homogenization	Oct-05–Oct-06	227.1	Municipal	Increase VSR destruction; increase biogas production	
Microsludge	Homogenization	Jul-07–Jul-08		Municipal		
Microsludge	Homogenization	2004 pilots	17.5–227		35 to 90% > VSR	3 Pilots
Crown	Pressure-Release Disintegration	May-02	15.6	Municipal	32% VSR Improvement	Operational
Crown	Pressure-Release Disintegration	Oct-02	3.6	Municipal	38.9% VSR Improvement	Operational
Crown	Pressure-Release Disintegration	Dec-02	24	Municipal	32% VSR Improvement	Operational

Manufacturer	Process	In Operation	Design Flow (MLD)	Municipal/Industrial	Range of Reported Impacts*	No. of Operational Installations
Crown	Pressure-Release Disintegration	May-03	42.6	Municipal	Carbon Sourcing role - not digestion enhancement. 50% reduction in activated sludge. Sludge volume index droped to 80 mg/L from 110 mg/L. No methanol dosing necessary.	Operational
Crown	Pressure-Release Disintegration	Nov-03	3.84	Municipal	20% VSR Improvement	Operational
Crown	Pressure-Release Disintegration	May-04	2.64	Municipal		Operational
Crown	Pressure-Release Disintegration	Jan-05		Industrial	Industrial sludge disintegration. Elimination of foaming due to filamentatious bacteria.	Operational
Crown	Pressure-Release Disintegration	Mar-05	3.6	Municipal	32	Operational
Crown	Pressure-Release Disintegration	Oct-05	2.16	Municipal	No data available as operational period is too short.	Operational
Crown	Pressure-Release Disintegration	Feb-06	11.16	Municipal	No data available because of installation of a new thickener.	Operational
Crown	Pressure-Release Disintegration	Apr-06	42	Municipal	System testing ongoing. Interim report shows a higher organic destruction of 12% and an improvement in dewatering of 3%.	Operational

Manufacturer	Process	In Operation	Design Flow (MLD)	Municipal/Industrial	Range of Reported Impacts*	No. of Operational Installations
Crown	Pressure-Release Disintegration	Nov-07	15.6	Municipal	Due to be commissioned in November 2007	Construction
Crown	Pressure-Release Disintegration	Mar-06	78	Municipal	22	Operational
Crown	Pressure-Release Disintegration	Sep-06	12	Municipal	The reduction in DS within the anaerobic digester has improved by 16 %. Polymer consumption down by 10%.	Operational
Crown	Pressure-Release Disintegration	Nov-06	102	Municipal	Carbon Sourcing role – Not digestion enhancement. Initial tests show that the organic nitrogen in the effluent is decreased by 3 mg/L	Operational
Crown	Pressure-Release Disintegration	Dec-07	15	Municipal	System test is finished in the End of 2007 no results available	Construction
Crown		Sep-03	90	Municipal		Operational
Crown		Jul-04	11.76	Municipal		Operational
Crown				Municipal		Pilot
Crown				Municipal		Pilot
Crown				Municipal		Pilot
Crown				industrial		Pilot
Crown				Municipal		Pilot
Crown				Municipal		Pilot

Manufacturer	Process	In Operation	Design Flow (MLD)	Municipal/Industrial	Range of Reported Impacts*	No. of Operational Installations
Crown	Pressure-Release Disintegration	2002	2.16–102			16 Operational; 2 Construction; 6 Pilot
OpenCel	Pulsed Electric Field			Municipal	24% less WAS	Pilot Test Completed
OpenCel	Pulsed Electric Field	Jul-07		Municipal	Too Early	Pilot Ongoing
OpenCel	Pulsed Electric Field			Municipal	24% less WAS	1 Pilot Operational, 1 Pilot Completed
Kady	Milling	1 Year	126.80	Municipal	i. Estimated net savings \$300,000 annually; ii. Concerns with settleability during peak runoff; iii. Sludge yield at least 12% lower than longer SRT	Full-Scale Trial
Kady	Milling	-		Municipal	i. Sludge yield reduced 33–43%; ii. Effluent BOD, COD, TSS and phosphorus removal improved; iii. Filamentous levels lower at older sludge ages (>10 days)	Pilot Trial

Manufacturer	Process	In Operation	Design Flow (MLD)	Municipal/Industrial	Range of Reported Impacts*	No. of Operational Installations
Kady	Milling	-		Municipal	i. Reduced wasting of activated sludge >50%; ii. Lower chemical dewatering costs; iii. Filamentous levels lower at older sludge ages (>10 days)	Pilot Trial
Kady		2003, 2-Month Test	4.90	Municipal		Pilot Trial
Kady		Lab		Industrial – Pulp and Paper		Laboratory Tests
Kady	Milling				12 to 50% less WAS	0 Operating, 4 Pilot, 1 Laboratory
Lysatec GMBh	Lysate Centrifuge			Municipal		
Lysatec GMBh	Lysate Centrifuge			Municipal		
Lysatec GMBh	Lysate Centrifuge			Municipal		
Lysatec GMBh	Lysate Centrifuge			Municipal		
Lysatec GMBh	Lysate Centrifuge			Municipal	15 – 26% > VSR	4 Operating

NOTES:
Population Equivalent (PE)= Flow-Equivalent Infl. Flow (For non-US plants, based on European standard of 120 liters/PE per day)
*The impact of the technologies as reported by the entity or the vendor associated with entity. WERF did not verify claims.

References

APHA. 1995. Standard methods. 19th Edition. American Public Health Association, Washington, D.C. ABS- Kady® International. 2006. Biolysis® Technical Bulletin. April.

Andreottoloa, G., P. Foladori, and M. Vian. 2006. An integrated approach for the evaluation of bacteria damage and biodegradability increase in sonication of excess sludge. *Proceedings of the International Water Association World Water Congress*. Beijing, China.

Atkins Ltd. 2005. *Report: Anaerobic Baffled Reactor (ABR) Pilot Plant Demonstration*. Prepared for Orange County Sanitation District. March.

Atkins Ltd. 2007. *Phase 3 of Anaerobic Baffled Reactor Study: Full-scale Retrofit Final Report*. Prepared for Orange County Sanitation District.

Banaszka, J.E. and A.M. Klein. 2007. Pulsed electric field treatment of waste activated sludge: Application to waste biosolids reduction and increased methane production. *Arizona Water & Pollution Control Association Annual Conference* (presentation). May.

Basu, S., M.Abu-Orf, M. Laquidara, O. Boe, G. Sanchez, G. Law, C. Muller, and J. Novak. 2004. High rate anaerobic digestion with mechanical shear – a case study at the Caldwell, ID WWTP. *Proceedings of the Water Environment Federation Annual Conference*. New Orleans, Louisiana.

Batstone, D.J., J. Keller, I. Angelidaki, S.V. Kalyuzhnyi, S.G. Pavlostathis, A. Rozzi, W.T.M. Sanders, H. Siegrist, and V.A. Vavilin. 2002. *Anaerobic Digestion Model No. 1 (ADM1)*. ISBN: 1900222787. IWA Publishing. London, UK.

Borgatti, D. 2004. Evaluation of the KADY® Biolysis® system at the 67-MGD Springfield Regional Wastewater Treatment Facility. Seminar presentation. University of Massachusetts.

Bradford, M.M., 1976. A rapid and sensitive method for the quantitation of microgram quantities of protein utilizing the principle of protein dye binding. Anal. Biochem. 72, 248-254.

CH2M HILL. 2005. *Fourth Quarterly Report, Project 2.2 Enhanced Energy Recovery through Optimization of Anaerobic Digestion and Microturbines – Commonwealth Energy Biogas/PV Mini-Grid Renewable Resources Program*. Prepared for the California Energy Commission Public Interest Energy Research Renewables Program.

Chauzy, J., D. Cretenot, A. Bausseron, and S. Deleris. 2007. Anaerobic digestion enhanced by thermal hydrolysis: first reference BioThelys® at Saumur, France. *Proceedings of the International Water Association Facing Sludge Diversities Conference*. Antalya, Turkey.

Christensson, M. and T. Welander. 2001. Decreasing the biosludge production by applying the low sludge production (LSP) process on municipal wastewater – a pilot study. Lund, Sweden. ANOX AB.

Cook, M. and B. Russel. 1994. Energy spilling reactions of *Streptococcus bovis* and resistance of its membrane to proton conductance. *Applied and Environmental Microbiology* 60: 1,942-1,948.

Dental, S., M. Abu-Orf, P. Griffin, and E. Hansen. 2000. *Investigating the Effects of Electrical Arc Pre-treatment of Biosolids*. Water Environment Research Foundation, Project 97-REM-4.

Dereix, M., W. Parker, and K. Kennedy. 2005. Steam-explosion pre-treatment for enhancing anaerobic digestion of municipal wastewater sludge. *Proceedings of the Water Environment Federation Annual Conference*. Washington, D.C.

Dubois, M., Gilles, K.A., Hamilton, J.K., Rebers, P.A., and Smith, F. 1956. Colorimetric method for determination of sugars and related substances. Anal. Chem. 28:350

Dytczak, M.A., K.L. Londry, H. Siegrist, and J.A. Oleszkiewicz. 2006. Extracellular polymers in partly ozonated return activated sludge: impact on flocculation and dewaterability. *Proceedings of the International Water Association World Water Congress*. Beijing, China.

Dytczak, M.A., K.L. Londry, H. Siegrist, and J.A. Oleszkiewicz. 2007. Ozonation Reduces Sludge Production and Improves Denitrification. *Water Res*. 41: 543-550.

Eskicioglu, C., R.L. Droste, and K.J. Kennedy. 2006. Performance of continuous flow anaerobic sludge digesters after microwave pre-treatment. *Proceedings of the Water Environment Federation Annual Conference*. Dallas, Texas.

Fabiyi, M., R. Novak, A. Ried, A. Wieland, R. Carpa, and A. Sandon. 2007. Sludge reduction using ozone induces lysis. *Proceedings of the International Water Association Facing Sludge Diversities Conference*. Antalya, Turkey.

Fjordside, C. 2002. Full scale experience of retrofitting thermal hydrolysis to an existing anaerobic digester for the digestion of waste activated sludge. *Proceedings of the Water Environment Federation Residuals and Biosolids Management Conference*. Austin, Texas.

Frølund, B., R. Palmgren, K. Keiding, and P.H. Nielsen. 1996. Extraction of Extracellular Polymers from Activated Sludge Using a Cation Exchange Resin. *Water Res*. 30 (8): 1,749–1,758.

Frolund, B., T. Griebe, and P. H. Nielsen. 1995. Enzymatic activity in the activated sludge floc matrix. Appl. Microbiol. Biotechnol. 43:755-761.

Garuti, G., M. Dohanyos, and A. Tilche. 1992. Anaerobic-Aerobic combined process for the treatment of sewage with nutrient removal – the Ananox® Process. *Water Science and Technology* 25 (7): 383-394.

Gaudy, A.F., P.Y. Yang, and A.W. Obayashi. 1971. Studies on the total oxidation of activated sludge with and without hydrolytic treatment. *Journal of Water Pollution Control Federation* 43 (1): 40–52.

Griffiths, M. 2001. *Use of high powered ultrasound in the activated sludge process*. New Zealand.

Hach Company, Loveland, CO – Product Information.

Heinzmann, B., and G. Engel. 2006. Two-stage high rate digestion and phosphorus recovery. *Proceedings of the International Water Association Sustainable Sludge Management Conference*. Moscow, Russia.

Higgins, M.J. and J.T. Novak. 1997. Characterization of exocellular protein and its role in bioflocculation. *Journal of Environmental Engineering* 123: 479-485.

Horng, R.Y., H. Shao, and M.C. Chang. 2006. Reduction sludge yield and enhance performance by an integration of anaerobic/aerobic and hydrolysis process from experimental study to full scale plant. *Proceedings of the International Water Association Sustainable Sludge Management Conference*. Moscow, Russia.

Jeganathan, J., G. Nakhla, and A. Bassi. 2006. Novel hybrid PBR-UASB system for the treatment of high strength pet food wastewater. *Proceedings of the Water Environment Federation Annual Conference*. Dallas, Texas.

Johnson, B.R. and Y. Shang. 2005. The Applications and Limitations of ADM 1 in Municipal Wastewater Solids Treatment. *Proceedings of the 77rd WEF Conference and Exhibition*. New Orleans, Louisiana.

Johnson, B.R., G.T. Daigger, and J.T. Novak. 2008. Biological Sludge Reduction Process Modeling With Asm Based Models. *Proceedings of the 77rd WEF Conference and Exhibition*. Chicago, Illinois.

Khanal, S.K., H. Isik, S. Sung, and J.H. van Leeuwen. 2006. Ultrasound conditioning of waste activated sludge for enhanced aerobic digestion. *Proceedings of the International Water Association World Water Congress*. Beijing, China.

Khanthongthip, P. 2010. Evaluation of the Cannibal® Process. PhD dissertation in preparation. Virginia Tech, Blacksburg, Virginia.

Kopp J., K. Panter, and W. Ewert. 2007. Technology development for the improvement of sludge digestion and dewatering by thermal hydrolysis. *Proceedings of the International Water Association Facing Sludge Diversities Conference*. Antalya, Turkey.

Kruger, R. and F. Hogan. 2005. Using Sonix™ to enhance anaerobic digestion: an overview from different trials and installations. *Proceedings of the Water Environment Federation Residuals and Biosolids Management Conference*. Nashville, Tennessee.

Kumar, N., J.T. Novak, and S.N. Murthy. 2006a. Efficient nitrogen removal and volatile solids reduction in sequential anaerobic-aerobic digestion process. *Proceedings of the Water Environment Federation Residuals and Biosolids Management Conference*. Cincinnati, Ohio.

Kumar, N., J.T. Novak, and S.N. Murthy. 2006b. Effect of secondary aerobic digestion on properties of anaerobic digested biosolids. *Proceedings of the Water Environment Federation Annual Conference*. Dallas, Texas.

Lapinski, J. and A. Tunnacliffe. 2003. Reduction of suspended biomass in municipal wastewater using bdelloid rotifers. *Water Research* 37: 2,027-2,034.

Le, M.S. 2006. Thermophilic biological pre-treatments for MADs. *Proceedings of the Advances in Technology for the Anaerobic Digestion of Municipal Sludge Conference*. Manchester, United Kingdom.

Le, M.S. 2007. Enzymatic hydrolysis – Sustainable sludge management. Presentation. February.

Le, M.S. and D. Harrison. 2006. Application of enzymic hydrolysis technology for VFA production and agricultural recycling of sludge. *Proceedings of the International Water Association Sustainable Sludge Management Conference*. Moscow, Russia.

Manterola G., I. Uriarte, and L. Sancho. 2006. Evaluation of operational parameters of sludge ozonation in industrial scale plant – effect on organic and nitrogenous compounds. *Proceedings of the International Water Association Sustainable Sludge Management Conference*. Moscow, Russia.

Mason, T.J. and A. Tiehm (Ed.). 2001. *Advances in Sonochemistry Volume 6 – Ultrasound in Environmental Protection*. Elsevier Science B.V.

McCarty, P.L. and D.P. Smith. 1986. Anaerobic waste-water treatment 4. *Environ. Sci. Technol.* 20 (12): 1,200-1,206.

Mesloh, W., K. Snider, J. Cook, and E. Garner. 2007. Rock Springs, Wyoming WWTP expansion using unique solids treatment and reduction process. *Proceedings of the Water Environment Federation Residuals and Biosolids Management Conference*. Denver, Colorado.

Middleberg, A.P.J. 1995. Process-scale disruption of microorganisms. *Biotechnology Advances*. 13 (3): 491-551.

Moon, H.M., J.C. Young, and Y.T. Cho. 2004. Evaluating the impact of anoxic sludge recycle on net sludge yield. *Proceedings of the Water Environment Federation Annual Conference*. New Orleans, Louisiana.

Muller, C.D., M. Abu-Orf, C.D. Blumenschein, and J.T. Novak. 2005. The impact of ultrasonic energy on mesophilic anaerobic digestion. *Proceedings of the Water Environment Federation Residuals and Biosolids Management Conference*. Nashville, Tennessee.

Neis, U. 2000. Ultrasound in water, wastewater and sludge treatment. *Water* 21 (April).

Higgins, M. J. and Novak, J.T. (1997a). The Effect of Cations on the Settling and Dewatering of Activated Sludges. Water Environ. Res., 69: 215.

Higgins, M.J. and Novak, J.T. (1997b). Characterization of Exocellular Protein and Its Role in Bioflocculation. J. Environ. Engr. ASCE, 123, 479-485.Novak, J., and C. Park. 2004. Chemical conditioning of sludge. *Wat. Sci. Tech.* 49 (10): 73-80.

Novak, J.T., C. Park, M. Higgins, Y.-C. Chen, R. Morton, D. Gary, R. Forbes, and Z. Erdal. 2007. WERF odor study phase III: impacts of the MicroSludge® process on odor causing compounds. *Proceedings of the Water Environment Federation Residuals and Biosolids Management Conference*. Denver, Colorado.

Novak, J.T., D.H. Chon, B.-A. Curtis, and M. Doyle. 2006a. Reduction of sludge generation using the Cannibal® process: mechanisms and performance. *Proceedings of the Water Environment Federation Residuals and Biosolids Management Conference*. Cincinnati, Ohio.

Novak, J.T., M.E. Sadler, and S.N. Murthy. 2003. Mechanisms of floc destruction during anaerobic and aerobic digestion and the effect on conditioning and dewatering of biosolids. *Water Res*. 37: 3,136-3,144.

Novak, J.T., N. Kumar, and S.N. Murthy. 2006b. Combined anaerobic/aerobic digestion of sewage sludge for enhanced volatile solids reduction and nitrogen removal. *Proceedings of the International Water Association World Water Congress*. Beijing, China.

Okey, R.W. and D.H. Stensel. 1993. Uncouplers and activated sludge – the impact on synthesis and respiration. *Toxicological and Environmental Chemistry* 40: 235-254.

Pandolfe, W.D. 1993. Cell disruption by homogenization. APV publication.

Panter, K., D. Jolis, O.E. Solheim, T. Seyffarth, T. Fjaergard, and G. Sorensen. 2006. THyPAD – from pilot to full scale application at Hamar WWTP. *Proceedings of the Water Environment Federation Annual Conference*. Dallas, Tennessee.

Park, C., M.M. Abu-Orf, and J.T. Novak. 2006. The digestibility of waste activated sludges. *Water Environ. Res.* 78 (1): 59-68.

Rabinowitz, B. and R. Stephenson. 2006. Effect of Microsludge® on anaerobic digester performance and residuals dewatering at LA County's JWPCP. *Proceedings of the Water Environment Federation Annual Conference*. Dallas, Texas.

Rickermann, R. and B. Johnson. 2007. Cannibal® as a cost effective biosolids alternative. *Proceedings of the Water Environment Federation Residuals and Biosolids Management Conference*. Denver, Colorado.

Rogalla, F., A. Field, J. Kolarik, and J. Bates. 2006. UASB operating experience at full scale. *Proceedings of the International Water Association Association Sustainable Sludge Management Conference*. Moscow, Russia.

Rouse, J., O. Burica, M. Strazar, and M. Levstek. 2006. A pilot-plant study of a moving-bed biofilm reactor system using PVA gel as a biocarrier for removals of organic carbon and nitroge. *Proceedings of the International Water Association Biofilm Systems VI Conference*. Amsterdam, Netherlands.

Rouse, J., O. Burica, M. Strazar, and M. Levstek. 2007. Elimination of excess organic sludge from moving-bed biofilm process. *Proceedings of the Water Environment Federation Residuals and Biosolids Management Conference*. Denver, Colorado.

Roxburgh, R., F. Soroushian, R. Seiger, and P. Burrowes. 2005. Ultrasound for improved solids management – how, who, when and why. *Proceedings of the Water Environment Federation Residuals and Biosolids Management Conference*. Nashville, Tennessee.

Roxburgh, R., R. Seiger, B. Johnson, B. Rabinowitz, S. Goodwin, G. Crawford, and G. Daigger. 2006. Sludge minimization technologies – doing more to get less. *Proceedings of the Water Environment Federation Annual Conference*. Dallas, Texas.

Rozich, A.F., R.J. Colvin, and C.D. Hahn. 2004. Design and Operation of a High Strength Organic Wastewater Treatment System to Approach Zero Net Sludge Production at a Specialty Chemical Plant. *Proceedings of Water Environment Federation and the Air and Waste Management Specialty Conference*, August 22-25 2004, Philadelphia, Pennsylvania. Copyright ©2004, Water Environment Federation, Alexandria, Virginia.

Rudd, T., R. Sterritt, and J. Lester. 1984. Complexation of heavy metal by extracellular polymers in the activated sludge process. *J. Water Poll Con. Res.* 56: 1,260-1,268.

Sandino, J., F. Rogalla, J. Londoño, R. Sadir, B. Sumner, J. Kolarik, and A. Field. 2004. Selected Experiences in the Applications of UASB Technology for Large-scale Municipal Advanced Wastewater Treatment Applications. *Proceedings of the Water Environment Federation Annual Conference*. New Orleans, Louisiana.

Sandino, J., R. Roxburgh, D. Whitlock, and L. Fillmore. 2008. Current State of the Practice of Sludge Reduction Technologies. *Proceedings of the WEF Conference: Residuals and Biosolids.*

Speece, R.E., S. Boonyakitsombut, M. Kim, N. Azbar, and P. Ursillo. 2005. Overview of anaerobic treatment: critical factors in successful anaerobic propionate metabolism. *Proceedings of the Water Environment Federation Annual Conference*. Washington, D.C.

Stensel, H.D. and S.E. Strand. 2004. *Evaluation of Feasibility of Methods to Minimize Biomass Production from Biotreatment*. Water Environment Research Foundation. Alexandria, Virginia, and IWA Publishing, London, U.K.

Stephenson, R., B. Rabinowitz, S. Laliberte, and P. Elson. 2005. Teaching an old digester new tricks: full-scale demonstration of the MicroSludge® process to liquefy municipal waste activated sludge. *Proceedings of the Water Environment Federation Residuals and Biosolids Management Conference*. Nashville, Tennessee.

STOWA. 2003a. Pulsed Power Technology Sheets. February.

STOWA. 2003b. Baker Process: Lysate-Thickening Centrifuge Sheets. February.

Strand, S.E., G.N. Harem, and H.D. Stensel. 1999. Activated sludge yield using chemical uncouplers. *Water Environ. Res*. 71 (4): 454-458.

Susarla, P., H. Van Rhijn, and S. Singleton. 2006. Full scale implementation and proving of cell lysis technology at North Shore City: part 3. *Proceedings of the New Zealand Water and Wastes Association Annual Conference*. Christchurch, New Zealand.

Tchobanoglous, G., F.L. Burton, and D.H. Stensel. 2003. *Wastewater Engineering Treatment and Reuse, Fourth Edition*. New York City, New York: Metcalf and Eddy, McGraw Hill.

Tunnacliffe, A.G., and J. Lapinski. 2006. Biological digestion of waste fluids with particles comminuting step. United States Patent, No. 7,153,430 B2. December 26.

van Loosdrecht, M.C.M., and M. Henze. 1999. Maintenance, endogenous respiration, lysis, decay and predation. *Water Sci. Technol.* 39 (1): 107-117.

Vera, M.A., K. Nickel, and U. Neis. 2005 Disintegration of sewage sludge for improved anaerobic disintegration. *Proceedings of the 57th Annual Conference of the Western Canada Water and Wastewater Association*. Saskatoon, Saskatchewan, Canada.

Vergine P., G. Mening, R. Canziani, E. Ficara, M. Fabiyi, R. Novak, A. Sandon, A. Bianchi, and G. Bergna. 2007. Partial ozonation of activated sludge to reduce excess sludge production: Evaluation of effects on biomass activity in a full scale demonstration test. *Proceedings of the International Water Association Biosolids Specialty Conference*. Moncton, New Brunswick, Canada.

Wujcik, W.J., A.F. Rozich, and C.D. Hahn. 2003. Design and start-up of an advanced thermophilic treatment system for high strength wastewater from a pharmaceutical plant. *Proceedings of the Water Environment Federation Annual Conference*. Los Angeles, California.

Zabranska, J., M. Dohanyos, P. Jenicek, J. Kutil, and J. Cejka. 2006. Mechanical and rapid thermal disintegration methods of enhancement of biogas production – full-scale applications. *Proceedings of the International Water Association Sustainable Sludge Management Conference*. Moscow, Russia.

Zhou, H. and D. Smith. 2000. Ozone mass transfer in water and wastewater treatment: experimental observations using a 2D laser particle dynamics analyzer. *Wat. Res*. 34 (3): 909-921.

WASTEWATER UTILITY

Alabama
Montgomery Water Works & Sanitary Sewer Board

Alaska
Anchorage Water & Wastewater Utility

Arizona
Avondale, City of
Glendale, City of, Utilities Department
Mesa, City of
Peoria, City of
Phoenix Water Services Dept.
Pima County Wastewater Management
Safford, City of
Tempe, City of

Arkansas
Little Rock Wastewater Utility

California
Central Contra Costa Sanitary District
Corona, City of
Crestline Sanitation District
Delta Diablo Sanitation District
Dublin San Ramon Services District
East Bay Dischargers Authority
East Bay Municipal Utility District
El Dorado Irrigation District
Fairfield-Suisun Sewer District
Fresno Department of Public Utilities
Inland Empire Utilities Agency
Irvine Ranch Water District
Las Gallinas Valley Sanitary District
Las Virgenes Municipal Water District
Livermore, City of
Los Angeles, City of
Los Angeles County, Sanitation Districts of
Napa Sanitation District
Novato Sanitary District
Orange County Sanitation District
Palo Alto, City of
Riverside, City of
Sacramento Regional County Sanitation District
San Diego Metropolitan Wastewater Department, City of
San Francisco, City & County of
San Jose, City of
Santa Barbara, City of
Santa Cruz, City of
Santa Rosa, City of
South Bayside System Authority
South Coast Water District
South Orange County Wastewater Authority
South Tahoe Public Utility District
Stege Sanitary District
Sunnyvale, City of
Union Sanitary District
West Valley Sanitation District

Colorado
Aurora, City of
Boulder, City of
Greeley, City of
Littleton/Englewood Water Pollution Control Plant
Metro Wastewater Reclamation District, Denver

Connecticut
Greater New Haven WPCA
Stamford, City of

District of Columbia
District of Columbia Water & Sewer Authority

Florida
Broward, County of
Fort Lauderdale, City of
Jacksonville Electric Authority (JEA)
Miami-Dade Water & Sewer Authority
Orange County Utilities Department
Pinellas, County of
Reedy Creek Improvement District
Seminole County Environmental Services
St. Petersburg, City of
Tallahassee, City of
Toho Water Authority
West Palm Beach, City of

Georgia
Atlanta Department of Watershed Management
Augusta, City of
Clayton County Water Authority
Cobb County Water System
Columbus Water Works
Fulton County
Gwinnett County Department of Public Utilities
Savannah, City of

Hawaii
Honolulu, City & County of

Idaho
Boise, City of

Illinois
Decatur, Sanitary District of
Greater Peoria Sanitary District
Kankakee River Metropolitan Agency
Metropolitan Water Reclamation District of Greater Chicago
Wheaton Sanitary District

Indiana
Jeffersonville, City of

Iowa
Ames, City of
Cedar Rapids Wastewater Facility
Des Moines, City of
Iowa City

Kansas
Johnson County Wastewater
Unified Government of Wyandotte County/ Kansas City, City of

Kentucky
Louisville & Jefferson County Metropolitan Sewer District
Sanitation District No. 1

Louisiana
Sewerage & Water Board of New Orleans

Maine
Bangor, City of
Portland Water District

Maryland
Anne Arundel County Bureau of Utility Operations
Howard County Bureau of Utilities
Washington Suburban Sanitary Commission

Massachusetts
Boston Water & Sewer Commission
Massachusetts Water Resources Authority (MWRA)
Upper Blackstone Water Pollution Abatement District

Michigan
Ann Arbor, City of
Detroit, City of
Holland Board of Public Works
Saginaw, City of
Wayne County Department of Environment
Wyoming, City of

Minnesota
Rochester, City of
Western Lake Superior Sanitary District

Missouri
Independence, City of
Kansas City Missouri Water Services Department
Little Blue Valley Sewer District
Metropolitan St. Louis Sewer District

Nebraska
Lincoln Wastewater & Solid Waste System

Nevada
Henderson, City of
Las Vegas, City of
Reno, City of

New Jersey
Bergen County Utilities Authority
Ocean County Utilities Authority

New York
New York City Department of Environmental Protection

North Carolina
Charlotte/Mecklenburg Utilities
Durham, City of
Metropolitan Sewerage District of Buncombe County
Orange Water & Sewer Authority
University of North Carolina, Chapel Hill

Ohio
Akron, City of
Butler County Department of Environmental Services
Columbus, City of
Metropolitan Sewer District of Greater Cincinnati
Montgomery, County of
Northeast Ohio Regional Sewer District
Summit, County of

Oklahoma
Oklahoma City Water & Wastewater Utility Department
Tulsa, City of

Oregon
Albany, City of
Clean Water Services
Eugene, City of
Gresham, City of
Portland, City of Bureau of Environmental Services
Lake Oswego, City of
Oak Lodge Sanitary District
Water Environment Services

Pennsylvania
Hemlock Municipal Sewer Cooperative (HMSC)
Philadelphia, City of
University Area Joint Authority

South Carolina
Charleston Water System
Mount Pleasant Waterworks & Sewer Commission
Spartanburg Water

Tennessee
Cleveland Utilities
Murfreesboro Water & Sewer Department
Nashville Metro Water Services

Texas
Austin, City of
Dallas Water Utilities
Denton, City of
El Paso Water Utilities

WERF SUBSCRIBERS

Fort Worth, City of
Houston, City of
San Antonio Water System
Trinity River Authority

Utah

Salt Lake City Corporation

Virginia

Alexandria Sanitation Authority
Arlington, County of
Fairfax, County of
Hampton Roads Sanitation District
Hanover, County of
Henrico, County of
Hopewell Regional Wastewater Treatment Facility
Loudoun Water
Lynchburg Regional Wastewater Treatment Plant
Prince William County Service Authority
Richmond, City of
Rivanna Water & Sewer Authority

Washington

Everett, City of
King County Department of Natural Resources
Seattle Public Utilities
Sunnyside, Port of
Yakima, City of

Wisconsin

Green Bay Metro Sewerage District
Kenosha Water Utility
Madison Metropolitan Sewerage District
Milwaukee Metropolitan Sewerage District
Racine, City of
Sheboygan Regional Wastewater Treatment
Wausau Water Works

Water Services Association of Australia

ACTEW Corporation
Barwon Water
Central Highlands Water
City West Water
Coliban Water Corporation
Cradle Mountain Water
Gippsland Water
Gladstone Area Water Board
Gold Coast Water
Gosford City Council
Hunter Water Corporation
Logan Water
Melbourne Water
Moreton Bay Water
Onstream
Power & Water Corporation
Queensland Urban Utilities
SEQ Water
South Australia Water Corporation
Sunshine Coast Water
Sydney Catchment Authority
Sydney Water
Unity Water
Wannon Regional Water Corporation
Watercare Services Limited (NZ)
Water Corporation
Western Water
Yarra Valley Water

Canada

Edmonton, City of/Edmonton Waste Management Centre of Excellence
Lethbridge, City of
Regina, City of, Saskatchewan
Toronto, City of, Ontario
Winnipeg, City of, Manitoba

STORMWATER UTILITY

California

Fresno Metropolitan Flood Control District
Los Angeles, City of, Department of Public Works
Monterey, City of
San Francisco, City & County of
Santa Rosa, City of
Sunnyvale, City of

Colorado

Aurora, City of
Boulder, City of

Florida

Orlando, City of

Iowa

Cedar Rapids Wastewater Facility
Des Moines, City of

Kansas

Lenexa, City of
Overland Park, City of

Kentucky

Louisville & Jefferson County Metropolitan Sewer District

Maine

Portland Water District

North Carolina

Charlotte, City of, Stormwater Services

Pennsylvania

Philadelphia, City of

Tennessee

Chattanooga Stormwater Management

Texas

Harris County Flood Control District, Texas

Washington

Bellevue Utilities Department
Seattle Public Utilities

STATE

Connecticut Department of Environmental Protection
Kansas Department of Health & Environment
New England Interstate Water Pollution Control Commission (NEIWPCC)
Ohio Environmental Protection Agency
Ohio River Valley Sanitation Commission
Urban Drainage & Flood Control District, CO

CORPORATE

ADS LLC
Advanced Data Mining International
AECOM
Alan Plummer & Associates
Alpine Technology Inc.
Aqua-Aerobic Systems Inc.
Aquateam–Norwegian Water Technology Centre A/S
ARCADIS
Associated Engineering
Bernardin Lochmueller & Associates
Black & Veatch
Blue Water Technologies, Inc.
Brown & Caldwell
Burgess & Niple, Ltd.
Burns & McDonnell
CABE Associates Inc.
The Cadmus Group
Camp Dresser & McKee Inc.
Carollo Engineers Inc.
Carpenter Environmental Associates Inc.
CET Engineering Services
CH2M HILL
CRA Infrastructure & Engineering
CONTECH Stormwater Solutions
D&B/Guarino Engineers, LLC
Damon S. Williams Associates, LLC
Ecovation
EMA Inc.
Environmental Operating Solutions, Inc.
Environ International Corporation
Fay, Spofford, & Thorndike Inc.
Freese & Nichols, Inc.
ftn Associates Inc.
Gannett Fleming Inc.
Garden & Associates, Ltd.
Geosyntec Consultants
GHD Inc.
Global Water Associates
Greeley and Hansen LLC
Hazen & Sawyer, P.C.
HDR Engineering Inc.
HNTB Corporation
Hydromantis Inc.
HydroQual Inc.
Infilco Degremont Inc.
Jason Consultants LLC Inc.
Jordan, Jones, & Goulding Inc.
KCI Technologies Inc.
Kelly & Weaver, P.C.
Kennedy/Jenks Consultants
Larry Walker Associates
LimnoTech Inc.
Lombardo Associates, Inc.
The Low Impact Development Center Inc.
Malcolm Pirnie Inc.
Material Matters, Inc.
McKim & Creed
MWH
NTL Alaska, Inc.
O'Brien & Gere Engineers Inc.
Odor & Corrosion Technology Consultants Inc.
Parametrix Inc.
Parsons
Post, Buckley, Schuh & Jernigan
Praxair, Inc.
RMC Water & Environment
Ross & Associates Ltd.
SAIC
Siemens Water Technologies
The Soap & Detergent Association
Smith & Loveless, Inc.
Southeast Environmental Engineering, LLC
Stone Environmental Inc.
Stratus Consulting Inc.
Synagro Technologies Inc.
Tetra Tech Inc.
Trojan Technologies Inc.
Trussell Technologies, Inc.
URS Corporation
Wallingford Software
Westin Engineering Inc.
Wright Water Engineers
Zoeller Pump Company

INDUSTRY

American Electric Power
American Water
Anglian Water Services, Ltd.
Chevron Energy Technology
The Coca-Cola Company
Dow Chemical Company
DuPont Company
Eastman Chemical Company
Eli Lilly & Company
InsinkErator
Johnson & Johnson
Merck & Company Inc.
Procter & Gamble Company
Suez Environnment
United Utilities North West (UUNW)
United Water Services LLC
Veolia Water North America

Note: List as of 4/20/10

WERF Board of Directors

WERF Research Council

IWA Publishing's authorised EU representative for General Product
Safety Regulations is Diane D'Arras, 15 rue Duret, 75116 Paris,
France, e-mail: safety@iwap.co.uk.

Printed and bound by CPI Group (UK) Ltd, Croydon, CR0 4YY
12/05/2026
02108868-0013